T0419924

NONLINEAR PHYSICAL SCIENCE

NONLINEAR PHYSICAL SCIENCE

Nonlinear Physical Science focuses on recent advances of fundamental theories and principles, analytical and symbolic approaches, as well as computational techniques in nonlinear physical science and nonlinear mathematics with engineering applications.

Topics of interest in *Nonlinear Physical Science* include but are not limited to:

- New findings and discoveries in nonlinear physics and mathematics
- Nonlinearity, complexity and mathematical structures in nonlinear physics
- Nonlinear phenomena and observations in nature and engineering
- Computational methods and theories in complex systems
- Lie group analysis, new theories and principles in mathematical modeling
- Stability, bifurcation, chaos and fractals in physical science and engineering
- Nonlinear chemical and biological physics
- Discontinuity, synchronization and natural complexity in the physical sciences

SERIES EDITORS

INTERNATIONAL ADVISORY BOARD

Anjan Biswas
Daniela Milovic
Matthew Edwards

Mathematical Theory of Dispersion-Managed Optical Solitons

With 23 figures

Authors

Anjan Biswas
Dept of Applied Mathematics
& Theoretical Physics
Delaware State University
1200 N Dupont Highway
Dover, DE 19901-2277, USA
E-mail: biswas.anjan@gmail.com

Daniela Milovic
Faculty of Electronic Engineering
Department of Telecommunications
University of Nis
Serbia
E-mail: dachavuk@gmail.com

Matthew Edwards
School of Arts and Sciences
Department of Physics
Alabama A & M University
Normal, AL-35762, USA
E-mail: matthew.edwards@aamu.edu

ISSN 1867-8440 e-ISSN 1867-8459
Nonlinear Physical Science

ISBN 978-7-04-018292-7
Higher Education Press, Beijing

ISBN 978-3-642-10219-6 e-ISBN 978-3-642-10220-2
Springer Heidelberg Dordrecht London New York

Library of Congress Control Number: 2009938163

Cover design: Frido Steinen-Broo, EStudio Calamar, Spain

Printed on acid-free paper

Springer is part of Springer Science+Business Media (www.springer.com)

Preface

The concept of dispersion-managed (DM) optical solitons was introduced in the early 1990s. The advent of such DM solitons has changed the world of optical solitons. In fact, they are governed by the dispersion-managed nonlinear Schrödinger's equations (DM-NLSE), unlike in the case of classical solitons which are governed by the pure nonlinear Schrödinger's equations (NLSE). It is to be noted that the pure NLSE is integrable by the classical method of Inverse Scattering Transform, while the case of DM-NLSE is not integrable. This leads to a lot of challenges and hindrances in studying the DM-NLSE. Many methods have been introduced in order to study the DM-NLSE. They are the variational principle, soliton perturbation theory, moment method as well as the asymptotic analysis. These methods of studying the DM solitons have introduced a wider picture in this area.

This book introduces and exposes the concept of DM solitons from scratch. Later the soliton perturbation theory and the variational principle are introduced to study the dynamics of pulses that propagate through optical fibers. The types of optical fibers that are studied in this book are the polarization preserving fibers, birefringent fibers and finally the case of multiple channels is also taken into consideration. The asymptotic analysis is used to study the quasi-linear pulses in optical fibers where along with the dispersion, the nonlinearity is also managed. Later, the Gabitov-Turitsyn equations are derived for these three types of optical fibers using the asymptotic analysis. Subsequently, higher order asymptotic analysis is carried out to derive the higher order Gabitov-Turitsyn equations for these types of optical fibers. Finally, the issue of optical crosstalk is touched upon to complete the discussion.

It needs to be noted that there are quite a few technical aspects that are skipped in this text. Those issues are the collision induced frequency and timing jitter along with the amplitude jitter. The other important issue that has been deliberately skipped is the issue of four-wave mixing. Finally, one other important issue of DM solitons that has not been touched upon is the aspect of soliton radiation. These issues are not yet exhaustively studied in the context of DM solitons and therefore requires further development before being incorporated in this text. Although there are a few papers that have been published in this context, a substantial amount of work is yet to be done to complete these chapters.

This book is organized as follows: Chapter 1 introduces the necessity and importance of studying the dispersion-managed optical solitons as opposed to the classical or conventional optical solitons. Chapter 2 introduces the technicalities of dispersion-managed optical solitons, the conserved quantities as well as the soliton perturbation theory. Finally this chapter ends with a brief introduction to the variational principle. Chapter 3 focuses on the polarization preserving fibers. Three types of pulses are studied in this chapter. They are Gaussian, super-Gaussian and super-sech pulses. The soliton parameter dynamics are derived. Finally the stochastic perturbation of optical solitons is studied with the aid of soliton perturbation theory. Chapters 4 and 5 deal with the birefringent fibers and multiple channels. Chapter 6 details out the aspect of optical crosstalk in both the linear as well as the nonlinear regime. Chapter 7 derives the Gabitov-Turitsyn equation for polarization-preserving fibers, birefringent fibers as well as in the case of multiple channels by using the technique of multiple-scale perturbation expansion. In Chapter 8, the issue of quasi-linear pulses are studied which another form of optical pulses that are studied where the nonlinearity is also managed in addition to the group velocity dispersion. Finally, in Chapter 9, the higher order asymptotic analysis is carried out to derive the higher-order Gabitov-Turitsyn equations that serves as an opening to the future research in this direction.

This book is primarily intended for graduate students at the Masters and Doctoral levels in Applied Mathematics, Applied Physics and Engineering. Also undergraduate students, with senior standing, in Physics and Engineering will benefit out of this book. The pre-requisite of this book is a knowledge of Partial and Stochastic Differential Equations, Perturbation Theory and Quantum Mechanics.

Anjan Biswas
Daniela Milovic
Matthew Edwards

Acknowledgements

The first author of the book, namely Anjan Biswas, is extremely grateful to his parents for all their unconditional love in his upbringing, blessings, education, support, encouragement and sacrifices throughout his life, till today. He also wants to thank his spouse for all the help. He is most grateful to his only 5-year-old son *Sonny* for providing him with all the fun, laughter, light and unforgettable moments throughout the period of writing this book. Without such intimate moments with his son, this book would have not been possible.

The work by the first author of the book, Anjan Biswas, was fully supported by NSF-CREST Grant No: HRD-0630388 and Army Research Office (ARO) along with the Air Force Office of Scientific Research (AFOSR) under the award number: W54428-RT-ISP and these supports are genuinely and sincerely appreciated. This author is extremely thankful to Prof. Fengshan Liu, the Chair of the Department of Applied Mathematics and Theoretical Physics for his constant encouragement and all the help of many kinds, including release time, he received since he joined Delaware State University. He is also thankful and grateful to the grant of Prof. Liu, numbered DAAD 19-03-1-0375, from which he received financial support in numerous occasions.The first author is also extremely thankful to Dr. Noureddine Melikechi,the Dean of College of Mathematics, Natural Sciences and Technology and Dr. Harry L. Williams, Provost and Vice President of Delaware State University who also helped him on several occasions with his research activities at Delaware State University.

The second author of the book namely Daniela Milovic, first and foremost, *thanks God* for all the blessings throughout her life and studies. She also expresses profound thanks to her parents Milorad and Zora Milović for their constant encouragement, unconditional love, selfless sacrifices, providing a warm, comfortable atmosphere in which she could think, write and live. She offers a special word of thanks to her only 10-year-old son Vukašin who brought unforgettable moments of joy and happiness into her life and inspired her to write this book.

The work of the second author Daniela Milovic was supported by the Army grant numbered DAAD 19-03-1-0375 and this support is genuinely and sincerely appreciated. The second author is also thankful to Prof. Fengshan Liu, the Chair of the Department of Applied Mathematics and Theoretical Physics for all his support during her visit to this Department in Spring 2008.

Contents

Chapter 1
Introduction

Dispersion-Managed optical solitons or DM solitons was first introduced in the early 1990s and since then it became a very attractive topic for optical communications. Dispersion-Management is very important in dense wavelength-division multiplexed (DWDM) systems. DM solitons allow the formation of ultra-long-haul Tera bit level optical networks working in all optical mode and maintaining optical transparency over vast geographical regions. The more wavelengths in use, the greater the need for dispersion-management. In high channel count systems, channel spacing is very close and requires continuous broadband management of dispersion and dispersion slope [1–10].

The discovery of dispersion-managed optical soliton has introduced plenty of new methods for high rate data transmission. By proper choice of parameters both Gordon-Haus effect and four-wave-mixing can be significantly suppressed thus providing nearly error-free transmission. It was shown in 1997 that bright solitons can exist in DM fibers and they can exist at normal path average dispersion. Dark solitons can also exist in such fibers, even at anomalous path averaged dispersion. This implies that the range of existence of bright and dark solitons overlap and therefore it becomes possible, for the first time, to analyze interaction between bright and dark solitons.

The propagation of DM solitons is governed by the dispersion-managed nonlinear Schrödinger's equation (DM-NLSE). Dispersion management forces each soliton to propagate in the normal dispersion regime of a fiber during every map period. When the map period is a fraction of the nonlinear length, the nonlinear effects become insignificant, leading to linear pulse evolution over the map period. If the self-phase modulation (SPM) effects are balanced by the average dispersion, solitons can survive in an average sense even on a longer length scale.

The NLSE that governs the classical or conventional soliton propagation is integrable by the classical method of Inverse Scattering Transform (IST), unlike DM-NLSE which is not integrable by IST, since the Painleve test of integrability [2] will fail. So, several methods have been introduced in order to study DM-NLSE that includes variational approach, collective variables approach, soliton perturbation theory, moment method as well as the asymptotic analysis. These methods really brought wider picture in the area of DM solitons.

The main feature of DM soliton is that it does not maintain its shape, width or peak power, unlike a fundamental soliton. However, DM soliton parameters repeat through dispersion map from period to period. This makes DM solitons applicable in communications in spite of changes in shape, width or peak power. From a systems standpoint, these DM solitons perform better.

By a proper choice of initial pulse energy, width and chirp will periodically propagate in the same dispersion map. The pulse energies much smaller than critical energy should be avoided in designing DM soliton system, whereas if the pulse energy is the same as the critical energy, it is the most suitable situation. An inappropriate choice of initial pulse energy may cause pulse interaction and thus lead to detrimental pulse distortion. The required map period becomes shorter as the bit rate increases.

The main difference between the average group velocity dispersion (GVD) solitons and DM solitons lies in its higher peak power requirements for sustaining DM solitons. The larger energy of DM solitons benefits a soliton system by improving the signal-to-noise ratio (SNR) and decreasing the timing jitter. The use of periodic dispersion map enables ultra high data transmission over large distances without using any in-line optical filters since the periodic use of dispersion-compensating fibers (DCF) reduces timing jitter by a large factor.

An important application of the dispersion-management is in upgrading the existing terrestrial networks employing standard fibers. Recent experiments show that the use of DM solitons has the potential of realizing transoceanic light-wave systems capable of operating with a capacity of 1 Tb/s or more.

Optical amplifiers compensate fiber losses but on the other hand induce timing jitter. This phenomena is mainly caused by the change of soliton frequency which affects the group velocity or the speed at which the pulse propagates through the fiber. The timing jitter can become an appreciable fraction of a bit slot for long-haul systems as bit slots become less than 100 ps. If left uncontrolled, such jitter can cause large power penalties. For DM solitons timing jitter is considerably smaller than that for fundamental solitons and the physical reason for jitter reduction is related to the enhanced energy of the DM solitons. From a practical point of view, reduced timing jitter of DM solitons permits much longer transmission distances.

As the bit rate increases, soliton-soliton interaction becomes a critical issue. Collision length depends on the details of dispersion map. The system performance can be optimized by an appropriate choice of map strength.

Dispersion-management can be efficiently used in several situations as follows:

1. For optimum pulse generation in a mode-locked laser operating at a wavelength around 1 μm or shorter. The normal chromatic dispersion has to be over compensated in order to utilize the anomalous dispersion regime, where soliton effects can help to obtain shorter pulses. It is usually also necessary to compensate carefully the higher-order dispersion, i.e., to control the group delay dispersion over a significant optical bandwidth.
2. In a mode-locked fiber laser, dispersive and nonlinear effects can become so strong that the pulse parameters (including the pulse duration and chirp) vary significantly during each resonator round-trip. With a suitable combination of fibers exhibiting normal and anomalous dispersion, a stretched-pulse fiber laser can be realized, which can generate pulses (DM solitons) with significantly higher pulse energy than with e.g. soliton mode locking.
3. Similar effect can be used in optical fiber communications: a fiber-optic link consisting of a periodic arrangement of fibers with normal and anomalous dispersion can help to suppress nonlinear effects such as channel crosstalk via four-wave mixing. It is possible to suppress the Gordon-Haus timing jitter at the same time, if the average chromatic dispersion is zero.

References

1. F. Abdullaev, S. Darmanyan & P. Khabibullaev. *Optical Solitons.* Springer, New York, NY. USA. (1993).
2. M. J. Ablowitz & H. Segur. *Solitons and the Inverse Scattering Transform.* SIAM. Philadelphia, PA. USA. (1981).
3. G. P. Agrawal. *Nonlinear Fiber Optics.* Academic Press, San Diego, CA. USA. (1995).
4. N. N. Akhmediev & A. Ankiewicz. *Solitons, Nonlinear Pulses and Beams.* Chapman and Hall, London. UK. (1997).
5. A. Biswas & S. Konar. *Introduction to Non-Kerr Law Optical Solitons.* CRC Press, Boca Raton, FL, USA. (2006).
6. A. Hasegawa & Y. Kodama. *Solitons in Optical Communications.* Clarendon Press, Oxford. UK. (1995).
7. Y. Kivshar & G. P. Agrawal. *Optical Solitons: From Fibers to Photonic Crystals.* Academic Press, San Diego, CA. USA. (2003).
8. B. A. Malomed. *Soliton Management in Periodic Systems.* Springer, Heidelberg. DE. (2006).
9. L. Mollenauer & J. P. Gordon. *Soliton in Optical Fibers: Fundamentals and Applications.* Academic Press, San Diego, CA. USA. (2006).
10. V. E. Zakharov & S. Wabnitz. *Optical Solitons: Theoretical Challenges and Industrial Perspectives.* Springer, Heidelberg. DE. (1999).

Chapter 2
Nonlinear Schrödinger's Equation

In this chapter the nonlinear Schrodinger's equation (NLSE) will be derived from the basic principles of Electromagnetic Theory. This equation will be then modified in presence of dispersion-management. The conserved quantities of this dispersion-managed NLSE (DM-NLSE) will be derived. The variational principle used for solving the DM-NLSE will be introduced. Finally, this chapter will end with a brief introduction to the soliton perturbation theory.

2.1 Derivation of NLSE

Pulse propagation through nonlinear and dispersive medium is governed by the NLSE that is derived from Maxwell's equation. Maxwell's equations arise in Electromagnetic Waves and are described in a medium that is assumed to be isotropic with no free charges (i.e., no plasma). NLSE can be modified for the case of plasma generation by adding terms to the NLSE that account for multi-photon absorption and plasma index change.

The starting point is the Maxwell's equations which takes the following forms, assuming no free charges:

$$\nabla \cdot \boldsymbol{D} = 0 \tag{2.1}$$

$$\nabla \cdot \boldsymbol{B} = 0 \tag{2.2}$$

$$\nabla \times \boldsymbol{E} = -\frac{\partial \boldsymbol{B}}{\partial t} \tag{2.3}$$

$$\nabla \times \boldsymbol{H} = \frac{\partial \boldsymbol{D}}{\partial t} \tag{2.4}$$

where $\boldsymbol{E}$, $\boldsymbol{H}$ are electrical and magnetic fields while $\boldsymbol{D}$, $\boldsymbol{B}$ are the respective flux densities and all vectors depend on three spatial coordinates and time

(t). The spatial and time dependence of vector fields is not shown explicitly to keep the notations simple. First of all, separate $\boldsymbol{D}$ into linear and nonlinear parts as

$$\boldsymbol{D} = \epsilon_0 \boldsymbol{E} + \boldsymbol{P} \tag{2.5}$$

where ϵ_0 is the permittivity of vacuum and $\boldsymbol{P}$ is the induced polarization. For optical fibers, the magnetic flux $(\boldsymbol{B})$, that is written as

$$\boldsymbol{B} = \mu_0 \boldsymbol{H} + \boldsymbol{M} \tag{2.6}$$

where μ_0 is the vacuum permeability, can be simplified by taking $\boldsymbol{M} = 0$ because silica glass has no magnetic effect. The second term in the right-hand side of Eq.(2.5) represents nonlinear polarization that accounts for all higher order effects, frequency generation, nonlinear absorption, nonlinear index changes and others. The first term in the right-hand side of Eq.(2.5) on the other hand accounts for linear response.

The wave equation can now be derived by taking the curl of Eq. (2.3) and using Eqs.(2.4)–(2.6) as

$$\nabla \times \nabla \times \boldsymbol{E} = -\frac{1}{c^2}\frac{\partial^2 \boldsymbol{E}}{\partial t^2} - \mu_0 \frac{\partial \boldsymbol{P}}{\partial t^2} \tag{2.7}$$

where the speed of light in a vacuum is defined as $c = \sqrt{\epsilon_0 \mu_0}$. By virtue of Eq.(2.5) and using the identity from Vector Calculus, Eq.(2.7) can be rewritten as

$$\nabla^2 \boldsymbol{E} + \frac{\epsilon_0}{c^2}\frac{\partial^2 \boldsymbol{E}}{\partial t^2} = -\frac{1}{c^2}\frac{\partial \boldsymbol{P}}{\partial t^2} \tag{2.8}$$

This is the nonlinear wave equation in the time domain. Assume that the polarization does not change during the propagation, so scalars can be used instead of vectors. The coupling between orthogonal polarization states is, in general, very small and can be neglected. Applying Fourier transform (FT) to $\boldsymbol{E}(\boldsymbol{r}, t)$ and $\boldsymbol{P}(\boldsymbol{r}, t)$:

$$\tilde{\boldsymbol{E}}(\boldsymbol{r}, \omega) = \int_{-\infty}^{\infty} \boldsymbol{E}(\boldsymbol{r}, t) e^{i\omega t} dt \tag{2.9}$$

$$\tilde{\boldsymbol{P}}(\boldsymbol{r}, \omega) = \int_{-\infty}^{\infty} \boldsymbol{P}(\boldsymbol{r}, t) e^{i\omega t} dt \tag{2.10}$$

where

$$\boldsymbol{P}(\boldsymbol{r}, t) = \epsilon_0 \int_{-\infty}^{\infty} \chi(\rho, t - t') \boldsymbol{E}(\boldsymbol{r}, t - t') dt' \tag{2.11}$$

The linear susceptibility χ is scalar for an isotropic medium such as silica glass. Equation (2.11) represents the delayed nature of temporal response which is a feature that has important implications for optical fiber communications through chromatic dispersion. By virtue of Eq.(2.11), Eq.(2.7) can be written in the frequency domain as

$$\nabla \times \nabla \times \tilde{\boldsymbol{E}} = -\epsilon(\boldsymbol{r}, \omega)\frac{\omega^2}{c^2}\tilde{\boldsymbol{E}} \tag{2.12}$$

Now, the dielectric constant is frequency dependent and is, in general, complex. It is written as

$$\epsilon(\boldsymbol{r}, \omega) = 1 + \hat{\chi}(\boldsymbol{r}, \omega) \tag{2.13}$$

and $\tilde{\chi}(\boldsymbol{r}, \omega)$ is the Fourier transform of $\chi(\boldsymbol{r}, t)$. Its real and imaginary parts are related as

$$\epsilon = \left(n + \frac{ic\lambda}{2\omega}\right)^2 \tag{2.14}$$

Here the refractive index n and the absorption coefficient α are both frequency dependent and related to $\tilde{\chi}$ as

$$n = \{1 + \Re(\tilde{\chi})\}^2 \qquad \alpha = \frac{\omega}{nc}\Im(\tilde{\chi}) \tag{2.15}$$

where $\Re$ and $\Im$ represents the real and imaginary parts respectively. Fiber dispersion (chromatic and material dispersion), related to the frequency dependence of n, is one of the most important factors that limit performance of the fiber-optic communication systems.

For fibers with low optical losses, Eq.(2.12) can be simplified by taking ϵ to be real and replacing it by n^2. Since $n(\boldsymbol{r}, \omega)$ does not depend on spatial coordinate in both the core and in the cladding of a step-index fiber, the following identity can be used:

$$\nabla \times \nabla \times \tilde{\boldsymbol{E}} \equiv \nabla(\nabla \cdot \tilde{\boldsymbol{E}}) - \nabla^2 \tilde{\boldsymbol{E}} \tag{2.16}$$

Equation (2.16) holds approximately as long as the index change occur over a length scale much longer than the wavelength. By using Eq.(2.16) in Eq.(2.12), it is possible to obtain

$$\nabla^2 \tilde{\boldsymbol{E}} + n^2(\omega) k_0^2 \tilde{\boldsymbol{E}} = 0 \tag{2.17}$$

where the wave number k_0 is defined as

$$k_0 = \frac{\omega}{c} = \frac{2\pi}{\lambda} \tag{2.18}$$

and λ is the vacuum wavelength of the optical field oscillating at the frequency ω. A specific solution of the wave equation (2.17) is called *optical mode* and it must satisfy certain boundary conditions and its spatial distribution remains while propagating along optical fiber. Single-mode fibers support only the HE_{11} mode, also known as the fundamental mode of the fiber. All higher order modes are cut off at the operating wavelength. Each frequency component of the optical field propagates through a single mode fiber as

$$\tilde{\boldsymbol{E}}(\boldsymbol{r}, \omega) = \hat{\boldsymbol{x}} F(x, y) \tilde{B}(0, \omega) e^{i\beta z} \tag{2.19}$$

where $\hat{\boldsymbol{x}}$ is the polarization unit vector, $B(0,\omega)$ is the Fourier transform of the initial amplitude, and β is the propagation constant. The field distribution of the fundamental fiber mode $F(x,y)$ can be approximated by the Gaussian distribution

$$E_x = Ae^{-\frac{\rho^2}{\omega^2}}e^{i\beta z} \tag{2.20}$$

where w is the *field radius* and is referred to as the *spot size*. It is determined by fitting the exact distribution to the Gaussian function. To force a fiber to support only fundamental mode, the V-number of a fiber is an indicative measure of how many higher-order modes can propagate through the fiber [19, 74]. It is defined as

$$V = \frac{2\pi}{\lambda}aNA \tag{2.21}$$

where $k = 2\pi/\lambda$ is the wavelength of the radiation or the wave number, a is the fiber core radius and NA is the numerical aperture relating to the refractive index-step between the core and the cladding. To force a fiber to only support the propagation of the lowest-order fundamental mode, the V-number must be such that $V \leq 2.405$ [19, 74]. Fiber becomes multi-mode when the core radius is increased, and thus the V-number becomes higher and the step-index remains unchanged. The spot size is for $1.2 < V < 2.4$ [74].

$$\frac{w}{a} \approx 0.65 + 1.619V^{-\frac{3}{2}} + 2.879V^{-6} \tag{2.22}$$

The effective core area $A_{\mathrm{eff}} = \pi w^2$ determines the light confinement in the core. Nonlinear effects become stronger in fibers with smaller values of the effective core area.

For the laser pulse containing many optical cycles, the optical signal traveling in the z direction can be expressed as the product of a complex envelope $A(z,t)$ and a carrier wave:

$$E(z,t) = A(z,t)e^{i\{\omega_0 t - \beta(\omega_0)z\}} \tag{2.23}$$

By taking the Fourier transform of the signal $A(z,t)e^{i\omega_0 t}$ into its frequency spectrum $A(z,\omega)$ at any plane z and propagate each frequency component forward a small distance dz, and then inverse Fourier-transform these components to temporal domain, it is possible to obtain signal envelope $A(z,t)$ at the plane $z + dz$:

$$A(z+dz,t) = \frac{1}{2\pi}\int_{-\infty}^{\infty} d\Delta\omega \int_{-\infty}^{\infty} A(z,t')e^{i\left\{\Delta\omega(t-t')\right\}}e^{-i\Delta\beta dz}dt' \tag{2.24}$$

where $\Delta\omega = \omega - \omega_0$, $\Delta\beta = \beta(\omega) - \beta(\omega_0)$. The propagation constant $\beta(\omega)$ has the weak nonlinear dispersive form as

$$\beta(\omega) = \beta_0 + \beta_1(\Delta\omega) + \frac{\beta_2}{2!}(\Delta\omega)^2 + \frac{\beta_3}{3!}(\Delta\omega)^3 \tag{2.25}$$

where $\beta_m = (d^m\beta/d\omega^m)_{\omega=\omega_0}$, with $\beta_1 = 1/v_g$ and v_g being the group velocity. The group velocity dispersion (GVD) coefficient β_2 is related to the dispersion parameter D by

$$D = \frac{d}{d\lambda}\left(\frac{1}{v_g}\right) = -\frac{2\pi c}{\lambda^2}\beta_2 \tag{2.26}$$

The coefficient β_3 is related to the dispersion slope S as

$$S = \left(\frac{2\pi c}{\lambda^2}\right)^2 \beta_3 + \frac{4\pi c}{\lambda^3}\beta_2 \tag{2.27}$$

By replacing $\Delta\omega$ by $i\left(\partial A(z,t)\right)/\partial t$ it is possible to calculate $\left(\partial A(z,t)\right)/\partial z$ from Eq.(2.24) as

$$\frac{\partial A}{\partial z} + \beta_1\frac{\partial A}{\partial t} + i\beta_2\frac{\partial^2 A}{\partial t^2} - \frac{\beta_3}{6}\frac{\partial^3 A}{\partial t^3} = 0 \tag{2.28}$$

This is the basic propagation equation that governs the evolution of pulses inside a single-mode fiber. In absence of dispersion, namely when $\beta_2 = \beta_3 = 0$, optical pulse propagates without change in shape such that $A(z,t) = A(0, t-\beta_1 z)$. Equation (2.28) can be extended to include self-phase modulation and cross-phase modulation effects by adding a nonlinear term. The resulting equation is known as the nonlinear Schrödinger's equation (NLSE) and has the form:

$$\begin{aligned} &\frac{\partial A}{\partial z} + \beta_1\frac{\partial A}{\partial t} + i\beta_2\frac{\partial^2 A}{\partial t^2} - \frac{\beta_3}{6}\frac{\partial^3 A}{\partial t^3} \\ &= \frac{\alpha}{2}A + i\left[\gamma\left|A\right|^2 A + \frac{i}{\omega_0}\frac{\partial}{\partial t}(\left|A\right|^2 A) - T_R A\frac{\partial}{\partial t}(\left|A\right|^2)\right] \end{aligned} \tag{2.29}$$

2.1.1 Limitations of conventional solitons

In fibers with constant GVD solitons can propagate over unlimited distances without any distortion. However, in reality, optical fibers have small but finite losses thus requiring amplification for long distance unregenerated transmission. Laser amplifiers such as erbium-doped fiber amplifiers (EDFA) are usually used where the amplified spontaneous emission (ASE) noise is added to the signal at each amplification. The noise modulates the soliton frequency randomly that leads to the random timing jitter through GVD of the fiber. This timing jitter is called the Gordon-Haus (GH) timing jitter [19]. The magnitude of the GH timing jitter (variance of the fluctuation of the arrival time of a pulse) is proportional to the GVD of the fiber when the pulse width is kept constant. Thus one can expect smaller timing jitter when the GVD is reduced. However, the GVD cannot be reduced arbitrarily because the soliton energy and consequently the signal-to-noise ratio (SIR) at the receiver will reduce as SIR is proportional to the fiber GVD. Thus, the GH timing jitter sets a fundamental limit to soliton transmission.

The translation from frequency fluctuations to timing jitter through fiber GVD is a major source of hindrance to performance enhancement of soliton systems. Besides the GH effect the interaction of solitons as well as the interaction of solitons with acoustic waves are reduced due to the smaller value of the GVD, but with a limit placed by the SIR requirement.

The nonlinear interaction between adjacent pulses through exponential tails of sech pulses leads to packing solitons with small separation. Separation with larger than five times the pulse width are usually necessary. Thus, the spectral width of the signal is considerably larger than that of modulation formats used in linear transmission, like the non-return to zero (NRZ) format, that leads to inefficiency in bandwidth utilization. This is an inherent drawback of the soliton transmission.

2.1.2 Dispersion-management

For fiber transmission systems operating in linear regimes, the total dispersion accumulated over the system should be made as close to zero as possible to avoid dispersive pulse broadening. When a dispersion-shifted fiber is used, where the fiber GVD is almost constantly equal to zero along the system, however, fiber nonlinearity which is small but non-negligible, severely degrades the performance of DWDM systems mainly due to the FWM-induced crosstalk between channels. The effect of FWM is also effective in single channel systems if the signal wavelength is almost coincident with the zero dispersion wavelength of the fiber. In this case the noise in the vicinity of the signal spectrum is amplified and the signal spectral width is broadened.

To reduce the nonlinearity induced performance degradation, nonzero dispersion fibers needs to be used as transmission fibers that requires some sort of dispersion compensation. Dispersion compensation can be achieved in a number of ways. One way is the pre- and post-compensation in lumped fashion by using dispersion-compensating fibers or one can use the chirped fiber gratings. A third way is by using the span-by-span compensation with amplifier spans constructed by positive- and negative-dispersion transmission fibers. Midpoint inversion of signal spectrum can also be used as dispersion compensation.

Nonlinear return-to zero (RZ) pulse transmission in dispersion-compensated systems that consist of alternating normal- and anomalous-dispersion fibers was first reported in 1992 [44]. An experimental demonstration of such system was first reported in 1995 [44]. A numerical study showing the periodically stationary soliton-like solution exist in such system was reported in 1995 [44]. Since then it has been widely recognized that soliton-like RZ pulses which are periodically stationary propagating through optical fibers with alternating sign of GVD is much more advantageous over conventional solitons in fibers with constant GVD. Such a periodically stationary solution is called a dispersion-managed (DM) soliton.

When there exists a nonzero residual averaged dispersion and the fiber nonlinearity is insignificant, the averaged pulse width is broadened with a rate determined by the averaged dispersion. The pulse is eventually dispersed out. Fiber nonlinearity can compensate for the pulse broadening if the chirp induced by the nonlinearity can counter balance the chirp induced by the accumulated dispersion. When the initial pulse amplitude is properly chosen, so that the balance between the nonlinearity and dispersion is achieved, the evolution of the pulse becomes periodically stationary and a DM soliton is thus created.

A DM soliton is a periodically stationary isolated pulse propagating in a fiber with alternating positive and negative dispersion. There is no explicit closed-form expression available for the DM soliton solution even for the simplest dispersion arrangement. This is contrary to the case of conventional solitons in fibers with constant dispersion where soliton solutions are explicitly given. Full numerical simulations of the NLSE with varying dispersion, is most accurate but time consuming and difficult to extract physical insights. Analytical approaches by means of which one can get physical insights of the fundamental nature of the DM solitons are strongly desired. In this chapter, the variational principle, that is the primary and most widely studied technique of attacking the analytical aspects of DM soliton, will be introduced. Also, the soliton perturbation theory that is used to study the other technical aspects of DM solitons like the collision-induced frequency and timing jitter, amplitude jitter, will be taken up later in this chapter. The other analytical techniques namely the multiple-scale perturbation method and the asymptotic analysis will be seen in the context of quasi-linear pulses in the later chapters.

2.1.3 Mathematical formulation

The relevant equation is the nonlinear Schrödinger's equation (NLSE) with damping and periodic amplification [3–6]:

$$iq_z + \frac{D(z)}{2} q_{tt} + |q|^2 q = -i\Gamma q + i\left[e^{\Gamma z_a} - 1\right] \sum_{n=1}^{N} \delta(z - nz_a) q \tag{2.30}$$

Here, Γ is the normalized loss coefficient, z_a is the normalized characteristic amplifier spacing, and z and t represent the normalized propagation distance and the normalized time, respectively, while q represents the wave profile expressed in the non-dimensional units.

Also, $D(z)$ is used to model strong dispersion management. Now, decompose the fiber dispersion $D(z)$ into two components namely a path-averaged constant value δ_a and a term representing the large rapid variation due to large local values of the dispersion [6]. Thus,

$$D(z) = \delta_a + \frac{1}{z_a}\Delta(\zeta) \tag{2.31}$$

where $\zeta = z/z_a$. The function $\Delta(\zeta)$ is taken to have average zero over an amplification period, namely

$$\langle \Delta \rangle = \frac{1}{z_a}\int_0^{z_a} \Delta\left(\frac{z}{z_a}\right) dz = 0 \tag{2.32}$$

so that, the path-averaged dispersion D will have an average δ_a, namely,

$$\langle D \rangle = \frac{1}{z_a}\int_0^{z_a} D(z)dz = \delta_a \tag{2.33}$$

The proportionality factor in front of $\Delta(\zeta)$, in Eq. (2.31), is chosen so that both δ_a and $\Delta(\zeta)$ are quantities of order one. In practical situations, dispersion management is often performed by concatenating together two or more sections of given length with different values of fiber dispersion. In the special case of a two-step map, it is convenient to write the dispersion map as a periodic extension of [6]

$$\Delta(\zeta) = \begin{cases} \Delta_1, & 0 \le |\zeta| < \frac{\theta}{2} \\ \Delta_2, & \frac{\theta}{2} \le |\zeta| < \frac{1}{2} \end{cases} \tag{2.34}$$

where Δ_1 and Δ_2 are given by

$$\Delta_1 = \frac{2s}{\theta} \tag{2.35}$$

$$\Delta_2 = -\frac{2s}{1-\theta} \tag{2.36}$$

with the map strength s defined as

$$s = \frac{\theta\Delta_1 - (1-\theta)\Delta_2}{4} \tag{2.37}$$

Conversely,

$$s = \frac{\Delta_1\Delta_2}{4(\Delta_2 - \Delta_1)} \tag{2.38}$$

and

$$\theta = \frac{\Delta_2}{\Delta_2 - \Delta_1} \tag{2.39}$$

A typical two-step dispersion map is shown in Figure 2.1.

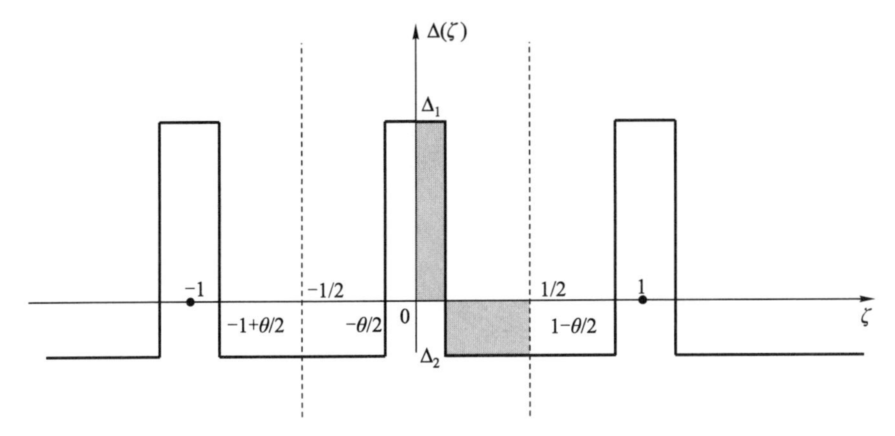

Fig. 2.1 Schematic diagram of a two-step dispersion map.

Taking into account the loss and amplification cycles by looking for a solution of Eq.(2.30) of the form $q(z,t) = P(z)u(z,t)$ for real P and taking P to satisfy

$$P_z + \Gamma P - \left[e^{\Gamma z_a} - 1\right] \sum_{n=1}^{N} \delta(z - nz_a)P = 0 \tag{2.40}$$

(1) transforms to

$$iu_z + \frac{D(z)}{2}u_{tt} + g(z)|u|^2 u = 0 \tag{2.41}$$

where

$$g(z) = P^2(z) = a_0^2 e^{-2\Gamma(z - nz_a)} \tag{2.42}$$

for $z \in [nz_a, (n+1)z_a)$ and $n > 0$ and also

$$a_0 = \left[\frac{2\Gamma z_a}{1 - e^{-2\Gamma z_a}}\right]^{\frac{1}{2}} \tag{2.43}$$

so that over each amplification period

$$\langle g(z) \rangle = \frac{1}{z_a} \int_0^{z_a} g(z)dz = 1 \tag{2.44}$$

Equation (2.41) is commonly known as the DM-NLSE and it governs the propagation of a dispersion-managed soliton through a polarization preserving optical fiber with damping and periodic amplification [3–6]. Figure 2.2 is profiles of DM solitons in the linear and logarithmic scales respectively.

In a polarization preserved optical fiber, it was seen in the last section that the propagation of solitons is governed by the scalar DM-NLSE given

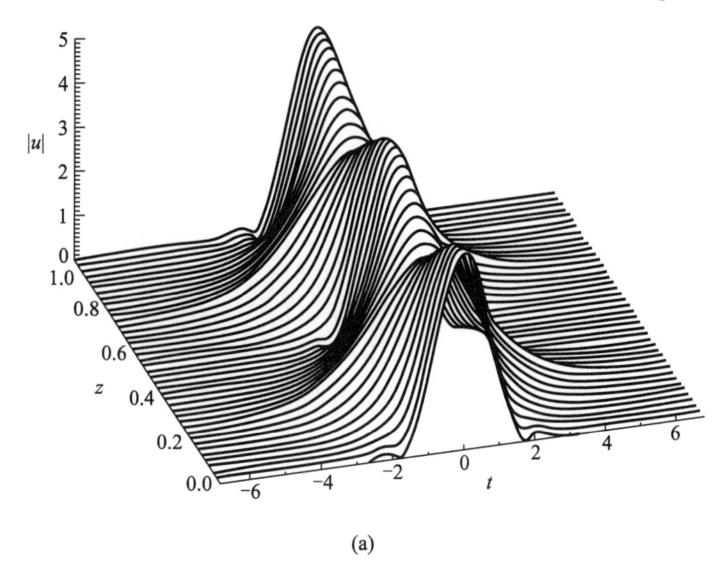

(a)

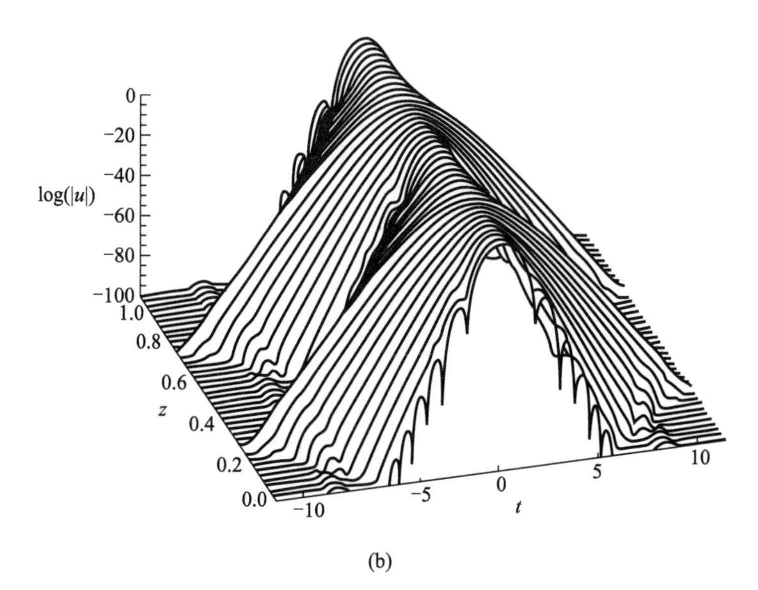

(b)

Fig. 2.2 Evolution of the pulse: (a) Linear scale, (b) logarithmic scale.

by Eq.(2.41). Note that in Eq.(2.41) if $D(z) = g(z) = 1$, NLSE is recovered. It is possible to integrate NLSE by the method of Inverse Scattering Transform (IST).

2.2 Integrals of motion

One of the intrinsic properties of DM-NLSE is that it has conserved quantities that is also known as *Integrals of Motion.* Conservation quantities are a common feature in Mathematical Physics, where they describe the conservation of fundamental physical quantities. In this section, the conserved quantities of DM-NLSE, given by Eq.(2.41), will be derived. Rewriting the NLSE in the form [21,37]

$$\frac{\partial T}{\partial z}+\frac{\partial X}{\partial t}=0 \tag{2.45}$$

it represents the *conservation law.* Here T is known as the *density* while X is known as the *flux.* Also neither the density nor the flux involve derivatives with respect to t. Thus, T and X may depend upon t, z, q, q_t, q_{tt}, ..., but not q_z. Now, if both T and X_t are integrable on $(-\infty,\infty)$, so that $X \to$ constant as $|t| \to \infty$, then Eq.(2.41) can be integrated to yield

$$\frac{d}{dz}\left(\int_{-\infty}^{\infty} T dt\right)=0 \tag{2.46}$$

so that

$$\int_{-\infty}^{\infty} T dt = constant \tag{2.47}$$

The integral of T, over all t, is therefore called the *constant of motion* or the *integral of motion.* For a dynamical system, with a finite number of degrees of freedom to be integrable, the system needs to have as many conserved quantities as the degrees of freedom. The first conserved quantity for the NLSE will now be derived.

Performing the operation (2.41) $\times q^*$ yields

$$iu^* q_z+\frac{D(z)}{2}u^* u_{tt}+g(z)\left|u\right|^4=0 \tag{2.48}$$

The complex conjugate of Eq.(2.48) is

$$-iuu_z^*+\frac{D(z)}{2}uu_{tt}^*+g(z)\left|u\right|^4=0 \tag{2.49}$$

Operating Eqs.(2.48)–(2.49) gives

$$i\left(u^* u_z+uu_z^*\right)+\frac{D(z)}{2}\left(u^* u_{tt}-uu_{tt}^*\right)=0 \tag{2.50}$$

which can be rewritten as

$$i\left(\left|u\right|^2\right)_z+\frac{D(z)}{2}\left(u^* u_t-uu_t^*\right)_t=0 \tag{2.51}$$

so that the flux is $u^* u_t - u u_t^*$. Integrating Eq.(2.51) with respect to t yields

$$\frac{d}{dz}\left(\int_{-\infty}^{\infty} |u|^2 \, dt\right) = 0 \tag{2.52}$$

since for a soliton u, u_t, u_{tt}, ... approach zero as $|t| \to \infty$, as mentioned before. Thus, the first conserved quantity for the NLSE is given by

$$E = \int_{-\infty}^{\infty} |u|^2 dt = constant \tag{2.53}$$

This conserved quantity is known as the *energy* or *wave power* while mathematically, it is the L_2 norm. Now, the second conserved quantity will be derived. The complex conjugate of Eq.(2.41) is given by

$$-iu_z^* + \frac{D(z)}{2} u_{tt}^* + g(z) |u|^2 u^* = 0 \tag{2.54}$$

Performing the operation $u_t^* \times$ (2.41) + $u_t \times$ (2.54) gives

$$i\left(u_z u_t^* - u_t u_z^*\right) + \frac{D(z)}{2}\left(u_t^* u_{tt} + u_t u_{tt}^*\right) + g(z)\left(u u_t^* + u^* u_t\right) |u|^2 = 0 \tag{2.55}$$

Now, performing the operation $q^* \times \frac{\partial}{\partial t}$(2.41) + $u \times \frac{\partial}{\partial t}$(2.54) yields

$$\begin{aligned} &i\left(u^* u_{tz} - u u_{tz}^*\right) + \frac{D(z)}{2}\left(u^* u_{ttt} + u u_{ttt}^*\right) \\ &+ g(z)\left[q^* \frac{\partial}{\partial t}\left\{\left(|q|^2\right) q\right\} + u \frac{\partial}{\partial t}\left\{\left(|u|^2\right) u^*\right\}\right] = 0 \end{aligned} \tag{2.56}$$

Then, Eqs.(2.55)–(2.56) leads to

$$\begin{aligned} &iD(z)\frac{\partial}{\partial z}\left(u u_t^* - u^* u_t\right) + \frac{D(z)}{2}\frac{\partial}{\partial t}\left(u_t u_t^*\right) \\ &- \frac{D(z)}{2}\left(u^* u_{ttt} + u u_{ttt}^*\right) + g(z)\left(|u|^2\right)_t |u|^2 \\ &- g(z)\left[\left(|u|^2\right)_t |u|^2 + 2|u|^2\left(|u|^2\right)_t\right] = 0 \end{aligned} \tag{2.57}$$

which simplifies to

$$\begin{aligned} &iD(z)\frac{\partial}{\partial z}\left(u u_t^* - u^* u_t\right) \\ &+ \frac{g(z)}{2}\frac{\partial}{\partial t}\left[2|u_t|^2 - \left(u^* u_{tt} + u u_{tt}^*\right) - 4|u|^2\right] = 0 \end{aligned} \tag{2.58}$$

Integrating Eq.(2.58) with respect to t gives

$$iD(z)\frac{d}{dz}\int_{-\infty}^{\infty}(u^*u_t - uu_t^*)dt = 0 \tag{2.59}$$

so that

$$M = iD(z)\int_{-\infty}^{\infty}(u^*u_t - uu_t^*)dt = constant \tag{2.60}$$

which is the second conserved quantity that is also known as the *linear momentum*. These are the only two conserved quantities that are known so far for the DM-NLSE. The third important quantity, namely the Hamiltonian (H), that is given by

$$H = \frac{1}{2}\int_{-\infty}^{\infty}\left(D(z)|u_t|^2 - g(z)|u|^4\right)dt \tag{2.61}$$

is not a conserved quantity. It can however be proved that H is an integral of motion of DM-NLSE if $D(z)$ and $g(z)$ are constants, in which case, it can actually be proved that there exists infinitely many conserved quantities for the NLSE given by Eq.(2.41). This is because when $D(z)$ and $g(z)$ are constants, it reduces to the case of classical or conventional solitons.

The conserved quantities will now be explicitly evaluated by assuming that the soliton pulses are given in the form [56]

$$\begin{aligned} u(z,t) = {} & A(z)f\left[B(z)\left\{t-\bar{t}(z)\right\}\right] \\ & \cdot \exp[iC(z)\left\{t-\bar{t}(z)\right\}^2 - i\kappa(z)\left\{t-\bar{t}(z)\right\} + i\theta(z)] \end{aligned} \tag{2.62}$$

where f represents the shape of the pulse. It could be a Gaussian type or an SG type pulse. Sometimes super-sech pulses are also considered. Here the parameters $A(z)$, $B(z)$, $C(z)$, $\kappa(z)$, $\bar{t}(z)$ and $\theta(z)$ respectively represent the soliton amplitude, the inverse width of the pulse, chirp, frequency, the center of the pulse and the phase of the pulse. It needs to be noted that, this approach is only approximate and does not account for characteristics such as energy loss due to continuum radiation, damping of the amplitude oscillations and changing of the pulse shape. For such a pulse form given by Eq.(2.62), the integrals of motion are

$$E = \int_{-\infty}^{\infty}|u|^2dt = \frac{A^2}{B}I_{0,2,0} \tag{2.63}$$

$$M = \frac{i}{2}D(z)\int_{-\infty}^{\infty}(u^*u_t - uu_t^*)dt = -\kappa D(z)\frac{A^2}{B}I_{0,2,0} \tag{2.64}$$

while the Hamiltonian is given by

$$H = \frac{1}{2}\int_{-\infty}^{\infty}\left(D(z)|u_t|^2 - g(z)|u|^4\right)dt$$

$$= \frac{D(z)}{2}\left(A^2 B I_{0,0,2} + 4\frac{A^2C^2}{B^3} I_{2,2,0} + \frac{\kappa^2 A^2}{B} I_{0,2,0}\right) - \frac{g(z)}{2}\frac{A^4}{B} I_{0,4,0} \qquad (2.65)$$

where the following notation was introduced:

$$I_{a,b,c} = \int_{-\infty}^{\infty} \tau^a f^b(\tau)\left(\frac{df}{d\tau}\right)^c d\tau \qquad (2.66)$$

for nonnegative integers a, b and c.

2.3 Soliton perturbation theory

The soliton parameters that were introduced in the previous subsection are now defined as [56]

$$\kappa(z) = \frac{i}{2}D(z)\frac{\int_{-\infty}^{\infty}(uu_t^* - u^*u_t)\,dt}{\int_{-\infty}^{\infty}|u|^2 dt} \qquad (2.67)$$

$$C(z) = \frac{i\int_{-\infty}^{\infty} t\,(uu_t^* - u^*u_t)\,dt}{\int_{-\infty}^{\infty} t^2|u|^2 dt} \qquad (2.68)$$

$$\overline{B}(z) = \frac{\int_{-\infty}^{\infty} t^2|u|^2 dt}{\int_{-\infty}^{\infty}|u|^2 dt} \qquad (2.69)$$

$$\overline{t}(z) = \frac{\int_{-\infty}^{\infty} t\,|u|^2\,dt}{\int_{-\infty}^{\infty}|u|^2 dt} \qquad (2.70)$$

where $\overline{B}$ is defined as the root-mean-square (RMS) width of the soliton. In case of DM solitons, the alternately varying dispersion as seen in Eqs.(2.31) and (2.34), makes the soliton breathe periodically. Thus, the soliton width does not stay constant although the energy of the soliton does. Hence, the RMS width is used in the analytical study of DM solitons. From these definitions, one can derive the variations of the soliton frequency, chirp, RMS-width as

$$\frac{d\kappa}{dz} = 0 \qquad (2.71)$$

$$\frac{dC}{dz} = D(z)\left(\frac{I_{0,0,2}}{I_{2,2,0}}B^4 - 12C^2 + \frac{I_{0,2,0}}{I_{2,2,0}}\kappa^2 B^2\right) - g(z)A^2B^2\frac{I_{0,4,0}}{I_{2,2,0}}. \qquad (2.72)$$

$$\frac{d\overline{B}}{dz} = -4D(z)\frac{C}{B}\frac{I_{2,2,0}}{I_{0,2,0}} \qquad (2.73)$$

Also, the velocity of the soliton is given by

$$v = \frac{d\bar{t}}{dz} = -\kappa D(z) \tag{2.74}$$

2.3.1 Perturbation terms

In presence of perturbation terms, the perturbed NLSE is given by

$$iu_z + \frac{D(z)}{2} u_{tt} + g(z)|u|^2 u = i\epsilon R\,[u, u^*] \tag{2.75}$$

where R represents the spatio-differential operator, although sometimes it could, very well, represent an integral operator. Also, the perturbation parameter ϵ is the relative width of the spectrum and $0 < \epsilon \ll 1$ by virtue of quasi-monochromaticity. In presence of these perturbation terms given by Eq.(2.75), the adiabatic variation of soliton parameters are given by [37, 56]

$$\frac{dE}{dz} = \epsilon \int_{\infty}^{\infty} (u^* R + uR^*)\, dt \tag{2.76}$$

From Eqs.(2.67)–(2.70), it is possible to obtain [56]

$$\begin{aligned} \frac{d\kappa}{dz} = \frac{\epsilon}{ED(z)} \bigg[& i \int_{-\infty}^{\infty} (u_t^* R - u_t R^*)\, dt \\ & - \kappa D(z) \int_{-\infty}^{\infty} (u^* R + uR^*)\, dx \bigg] \end{aligned} \tag{2.77}$$

$$\begin{aligned} \frac{dC}{dz} = {} & D(z) \left(\frac{I_{0,0,2}}{I_{2,2,0}} B^4 - 12C^2 + \frac{I_{0,2,0}}{I_{2,2,0}} \kappa^2 B^2 \right) \\ & - g(z) A^2 B^2 \frac{I_{0,4,0}}{I_{2,2,0}} + \frac{i\epsilon}{I_{2,2,0}} \frac{B^3}{A^2} \left[2 \int_{-\infty}^{\infty} t\,(u_t^* R - u_t R^*)\, dt \right. \\ & \left. + \int_{-\infty}^{\infty} (u^* R - uR^*)\, dx \right] + \frac{4\epsilon}{I_{2,2,0}} \frac{CB^3}{A^2} \int_{-\infty}^{\infty} t^2 (u^* R + uR^*)\, dx \end{aligned} \tag{2.78}$$

$$\frac{d\overline{B}}{dz} = -4D(z) \frac{C}{B} \frac{I_{2,2,0}}{I_{0,2,0}} + \frac{\epsilon}{E} \int_{-\infty}^{\infty} t^2 (u^* R + uR^*)\, dx \tag{2.79}$$

$$v = \frac{d\bar{t}}{dz} = -\kappa D(z) + \frac{\epsilon}{E} \int_{-\infty}^{\infty} t\,(u^* R + uR^*)\, dt \tag{2.80}$$

2.4 Variational principle

For a finite dimensional problem of a single particle, the temporal development of its position is given by the Hamilton's principle of least action [22, 37]. It states that the action given by the time integral of the Lagrangian is an extremum, namely,

$$\delta \int_{t_1}^{t_2} L(x, \dot{x}) dt = 0 \tag{2.81}$$

where x is the position of the particle and $\dot{x} = dx/dt$. The variational problem (2.81) then leads to the familiar Euler-Lagrange's (EL) equation [37]:

$$\frac{\partial L}{\partial x} - \frac{d}{dt}\left(\frac{\partial L}{\partial \dot{x}}\right) = 0 \tag{2.82}$$

Here, for Eq.(2.41), the Lagrangian is given by

$$L = \frac{1}{2}\int_{-\infty}^{\infty} \left[i\left(u^* u_z - u u_z^*\right) - D(z)|u_t|^2 + g(z)|u|^4 \right] dt \tag{2.83}$$

Now, using Eq.(2.62), the Lagrangian given by Eq.(2.83), reduces to

$$\begin{aligned} L = &-\frac{D(z)A^2}{2B^3}\left(B^4 I_{0,0,2} + 4C^2 I_{2,2,0} + \kappa^2 B^2 I_{0,2,0}\right) \\ &+ \frac{g(z)A^4}{2B} I_{0,4,0} - \frac{A^2}{B^3} I_{2,2,0}\frac{dC}{dz} + \frac{A^2}{B} I_{0,2,0}\left(\bar{t}\frac{d\kappa}{dz} - \frac{d\theta}{dz}\right) \end{aligned} \tag{2.84}$$

By the principle of least action, namely Eq.(2.81), the EL equation is [16]

$$\frac{\partial L}{\partial p} - \frac{d}{dz}\left(\frac{\partial L}{\partial p_z}\right) = 0 \tag{2.85}$$

where p is one of the six soliton parameters. Substituting $A, B, C, \kappa, \bar{t}$ and θ for p in Eq.(2.85) the following set of equations are obtained:

$$\frac{dA}{dz} = -ACD(z) \tag{2.86}$$

$$\frac{dB}{dz} = -2BCD(z) \tag{2.87}$$

$$\frac{dC}{dz} = \left(\frac{B^4}{2}\frac{I_{0,0,2}}{I_{2,2,0}} - 2C^2\right) D(z) - \frac{g(z)A^2B^2}{4}\frac{I_{0,4,0}}{I_{2,2,0}} \tag{2.88}$$

$$\frac{d\kappa}{dz} = 0 \tag{2.89}$$

$$\frac{d\bar{t}}{dz} = -\kappa D(z) \tag{2.90}$$

$$\frac{d\theta}{dz} = \left(\frac{\kappa^2}{2} - \frac{I_{0,0,2}}{I_{0,2,0}}B^2\right) D(z) + \frac{5g(z)A^2}{4}\frac{I_{0,4,0}}{I_{0,2,0}} \tag{2.91}$$

This Dynamical System of soliton parameters can be further analyzed based on the particular type of pulses that will be studied in the subsequent chapters. Furthermore, perturbation terms will be added and the adiabatic parameter dynamics of soliton parameters can also be obtained.

2.4.1 Perturbation terms

The perturbed DM-NLSE is going to be studied in this subsection. The adiabatic parameter dynamics of solitons is going to be studied in this subsection by the aid of variational principle. The perturbed DM-NLSE that is given by

$$iu_z + \frac{D(z)}{2}u_{tt} + g(z)|u|^2 u = i\epsilon R[u, u^*] \tag{2.92}$$

Here, once again, R represents the perturbation terms and ϵ is the perturbation parameter. In presence of the perturbation terms the EL equation modify to [27]

$$\frac{\partial L}{\partial p} - \frac{d}{dz}\left(\frac{\partial L}{\partial p_z}\right) = i\epsilon \int_{-\infty}^{\infty} \left(R\frac{\partial u^*}{\partial p} - R^*\frac{\partial u}{\partial p}\right) dt \tag{2.93}$$

where p represents the six soliton parameters. Once again, substituting A, B, C, κ, $\bar{t}$ and θ for p in Eq.(2.93) the following adiabatic evolution equations are obtained:

$$\frac{dA}{dz} = -ACD - \frac{\epsilon}{2I_{0,2,0}I_{2,2,0}} \int_{-\infty}^{\infty} \Re[Re^{-i\phi}](I_{0,2,0}\tau^2 - 3I_{2,2,0})f(\tau)d\tau \tag{2.94}$$

$$\frac{dB}{dz} = -2BCD - \frac{\epsilon B}{AI_{0,2,0}I_{2,2,0}} \int_{-\infty}^{\infty} \Re[Re^{-i\phi}](I_{0,2,0}\tau^2 - I_{2,2,0})f(\tau)d\tau \tag{2.95}$$

$$\frac{dC}{dz} = \left(\frac{B^4}{2}\frac{I_{0,0,2}}{I_{2,2,0}} - 2C^2\right) D - \frac{gA^2B^2}{4}\frac{I_{0,4,0}}{I_{2,2,0}} - \frac{\epsilon B^2}{2AI_{2,2,0}} \int_{-\infty}^{\infty} \Im[Re^{-i\phi}]\left(f(\tau) + 2\tau\frac{df}{d\tau}\right) d\tau \tag{2.96}$$

$$\frac{d\kappa}{dz} = \frac{2\epsilon}{ABI_{0,2,0}} \int_{-\infty}^{\infty} \left\{B^2\Im[Re^{-i\phi}]\frac{df}{d\tau} - 2C\Re[Re^{-i\phi}]\tau f(\tau)\right\} d\tau \tag{2.97}$$

$$\frac{d\bar{t}}{dz} = -\kappa D + \frac{2\epsilon}{ABI_{0,2,0}} \int_{-\infty}^{\infty} \Re[Re^{-i\phi}]\tau f(\tau) d\tau \tag{2.98}$$

$$\begin{aligned}\frac{d\theta}{dz} &= \left(\frac{\kappa^2}{2} - \frac{I_{0,0,2}}{I_{0,2,0}} B^2\right) D + \frac{5gA^2}{4} \frac{I_{0,4,0}}{I_{0,2,0}} \\ &+ \frac{\epsilon}{2ABI_{0,2,0}} \int_{-\infty}^{\infty} \left\{ B\Im[Re^{-i\phi}] \left(3f(\tau) + 2\tau \frac{df}{d\tau} \right) \right. \\ &\left. + 4\kappa \Re[Re^{-i\phi}]\tau f(\tau) \right\} d\tau \end{aligned} \tag{2.99}$$

where the notations

$$\tau = B(z)\left(t - \bar{t}(z)\right) \tag{2.100}$$

and

$$\phi = C(z)\left\{t - \bar{t}(z)\right\}^2 - \kappa(z)\left\{t - \bar{t}(z)\right\} + \theta(z) \tag{2.101}$$

was used. Also, $\Re$ and $\Im$ represent the real and imaginary parts, respectively. Note that Eqs.(2.86)–(2.91) are special cases of Eqs.(2.94)–(2.99) respectively for $\epsilon = 0$. These equations can be modified based on a particular type of pulse that is considered in optical fibers. This will be studied in detail in the following chapter.

References

1. F. Abdullaev, S. Darmanyan & P. Khabibullaev. *Optical Solitons.* Springer, New York, NY. USA. (1993).
2. M. J. Ablowitz & H. Segur. *Solitons and the Inverse Scattering Transform.* SIAM. Philadelphia, PA. USA. (1981).
3. M. J. Ablowitz, G. Biondini, S. Chakravarti, R. B. Jenkins & J. R. Sauer. "Four-wave mixing in wavelength-division-multiplexed soliton systems: damping and amplification". *Optics Letters.* Vol 21, No 20, 1646-1648. (1996).
4. M. J. Ablowitz, B. M. Herbst & C. M. Schober. "The nonlinear Schrödinger's equation; Asymmetric perturbations traveling waves and chaotic structures". *Mathematics and Computers in Simulation.* Vol 43, 3-12. (1997).
5. M. J. Ablowitz, G. Biondini, S. Chakravarty, R. B. Jenkins & J. R. Sauer. "Four-wave mixing in wavelength-division multiplexed soliton systems—Ideal fibers". *Journal of Optical Society of America B.* Vol 14, 1788-1794. (1997).
6. M. J. Ablowitz & G. Biondini. "Multiscale pulse dynamics in communication systems with strong dispersion management". *Optics Letters.* Vol 23, No 21, 1668-1670. (1998).
7. M. J. Ablowitz & P. A. Clarkson. "Solitons and symmetries". *Journal of Engineering Mathematics.* Vol 36, 1-9. (1999).
8. M. J. Ablowitz & T. Hirooka. "Resonant nonlinear interactions in strongly dispersion-managed transmission systems". *Optics Letters.* Vol 25, Issue 24, 1750-1752. (2000).
9. M. J. Ablowitz, G. Biondini & L. A. Ostrovsky. "Optical solitons: Perspectives and applications". *Chaos.* Vol 10, No 3, 471-474. (2000).
10. M. J. Ablowitz & T. Hirooka. "Nonlinear effects in quasi-linear dispersion-managed systems". *IEEE Photonics Technology Letters.* Vol 13, No 10, 1082-1084. (2001).

11. M. J. Ablowitz, T. Hirooka & G. Biondini. "Quasi-linear optical pulses in strongly dispersion-managed transmission system". *Optics Letters.* Vol 26, No 7, 459-461. (2001).
12. M. J. Ablowitz, G. Biondini & E. S. Olson. "Incomplete collisions of wavelength-division multiplexed dispersion-managed solitons". *Journal of Optical Society of America B.* Vol 18, No 3, 577-583. (2001).
13. M. J. Ablowitz & T. Hirooka. "Nonlinear effects in quasilinear dispersion-managed pulse transmission". *IEEE Journal of Photonics Technology Letters.* Vol 26, 1846-1848. (2001).
14. M. J. Ablowitz & T. Hirooka. "Intrachannel pulse interactions and timing shifts in strongly dispersion-managed transmission systems". *Optics Letters.* Vol 26, No 23, 1846-1848. (2001).
15. M. J. Ablowitz & T. Hirooka. "Intrachannel pulse interactions in dispersion-managed transmission systems: energy transfer". *Optics Letters.* Vol 27, No 3, 203-205. (2002).
16. M. J. Ablowitz & T. Hirooka. "Managing nonlinearity in strongly dispersion-managed optical pulse transmission". *Journal of Optical Society of America B.* Vol 19, No 3, 425-439. (2002).
17. M. J. Ablowitz, G. Biondini, A. Biswas, A. Docherty, T. Hirooka & S. Chakravarty. "Collision-induced timing shifts in dispersion-managed soliton systems". *Optics Letters.* Vol 27, 318-320. (2002).
18. M. J. Ablowitz, T. Hirooka & T. Inoue. "Higher-order asymptotic analysis of dispersion-managed transmission systems: solutions and their characteristics". *Journal of Optical Society of America B.* Vol 19, No 12, 2876-2885. (2002).
19. G. P. Agrawal. *Nonlinear Fiber Optics.* Academic Press, San Deigo, CA. USA. (1995).
20. C. D. Ahrens, M. J. Ablowitz, A. Docherty, V. Oleg, V. Sinkin, V. Gregorian & C. R. Menyuk. "Asymptotic analysis of collision-induced timing shifts in return-to-zero quasi-linear systems with pre- and post-dispersion compensation". *Optics Letters.* Vol 30, 2056-2058. (2005).
21. N. N. Akhmediev & A. Ankiewicz. *Solitons, Nonlinear Pulses and Beams.* Chapman and Hall. (1997).
22. D. Anderson. "Variational approach to nonlinear pulse propagation in optical fibers". *Physical Review A.* Vol 27, No 6, 3135-3145. (1983).
23. A. Biswas. "Dynamics of Gaussian and super-Gaussian solitons in birefringent optical fibers". *Progress in Electromagnetics Research.* Vol 33, 119-139. (2001).
24. A. Biswas. "Dispersion-Managed vector solitons in optical fibers". *Fiber and Integrated Optics.* Vol 20, No 5, 503-515. (2001).
25. A. Biswas. "Gaussian solitons in birefringent optical fibers". *International Journal of Pure and Applied Mathematics.* Vol 2, No 1, 87-104. (2002).
26. A. Biswas. "Super-Gaussian solitons in optical fibers". *Fiber and Integrated Optics.* Vol 21, No 2, 115-124. (2002).
27. A. Biswas. "Dispersion-managed solitons in optical fibers" *Journal of Optics A.* Vol 4, No 1, 84-97. (2002).
28. A. Biswas. "Integro-differential perturbation of dispersion-managed solitons". *Journal of Electromagnetic Waves and Applications.* Vol 17, No 4, 641-665. (2003).
29. A. Biswas. "Dispersion-managed solitons in optical couplers". *Journal of Nonlinear Optical Physics and Materials.* Vol 12, No 1, 45-74. (2003).
30. A. Biswas. "Gabitov-Turitsyn equation for solitons in multiple channels". *Journal of Electromagnetic Waves and Applications.* Vol 17, No 11, 1539-1560. (2003).
31. A. Biswas. "Gabitov-Turitsyn equation for solitons in optical fibers". *Journal of Nonlinear Optical Physics and Materials.* Vol 12, No 1, 17-37. (2003).
32. A. Biswas. "Dispersion-Managed solitons in multiple channels". *Journal of Nonlinear Optical Physics and Materials.* Vol 13, No 1, 81-102. (2004).
33. A. Biswas. "Theory of quasi-linear pulses in optical fibers". *Optical Fiber Technology.* Vol 10, No 3, 232-259. (2004).

34. A. Biswas. "Perturbations of dispersion-managed optical solitons". *Progress in Electromagnetics Research.* Vol 48, 85-123. (2004).
35. A. Biswas. "Dispersion-Managed solitons in birefringent fibers and multiple channels". *International Mathematical Journal.* Vol 5, No 5, 477-515. (2004).
36. A. Biswas. "Stochastic perturbation of dispersion-managed optical solitons". *Optical and Quantum Electronics.* Vol 37, No 7, 649-659. (2005).
37. A. Biswas & S. Konar. *Introduction to Non-Kerr Law Optical Solitons.* CRC Press, Boca Raton, FL. USA. (2006).
38. D. Breuer, F. Kuppers, A. Mattheus, E. G. Shapiro, I. Gabitov & S. K. Turitsyn. "Symmetrical dispersion compensation for standard monomode fiber based communication systems with large amplifier spacing". *Optics Letters.* Vol 22, No 13, 982-984. (1997).
39. D. Breuer, K. Jürgensen, F. Küppers, A. Mattheus, I. Gabitov & S. K. Turitsyn. "Optimal schemes for dispersion compensation of standard monomode fiber based links". *Optics Communications.* Vol 40, No 1-3, 15-18. (1997).
40. V. Cautaerts, A. Maruta & Y. Kodama. "On the dispersion-managed soliton". *Chaos.* Vol 10, No 3, 515-528.
41. F. J. Diaz-Otero & P. Chamorro-Posada. "Interchannel soliton collisions in periodic dispersion maps in the presence of third order dispersion". *Journal of Nonlinear Mathematical Physics.* Vol 15, Supplement 3, 137-143. (2008).
42. S. L. Doty, J. W. Haus, Y. J. Oh & R. L. Fork. "Soliton interaction on dual core fibers". *Physical Review E.* Vol 51, No 1, 709-717. (1995).
43. A. Ehrhardt, M. Eiselt, G. Goßkopf, L. Küller, W. Pieper, R. Schnabel, G. H. Weber. "Semiconductor laser amplifier as optical switching gate". *Journal of Lightwave Technology.* Vol 11, No 8, 1287-1295. (1993).
44. A. Hasegawa. "A historical review of applications of optical solitons for high speed communications". *Chaos.* Vol 10, No 3, 475-485. (2000).
45. A-H. Guan & Y-H. Wang. "Experimental study of interband and intraband crosstalk in WDM networks". *Optoelectronics Letters.* Vol 4, No 1, 42-44. (2008).
46. E. Iannone, R. Sabella, M. Avattaneo & G. De Paolis. "Modeling of in-band crosstalk in WDM optical networks". *Journal of lightwave technology.* Vol 17, No 7, 1135-1141. (1999).
47. W. L. Kath & N. F. Smyth. "Soliton evolution and radiation loss for the nonlinear Schrodinger's equation". *Physical Review E.* Vol 51, No 2, 1484-1492. (1995).
48. C. M. Khalique & A. Biswas. "A Lie Symmetry approach to nonlinear Schrödinger's equation with non-Kerr law nonlinearity". *Communications in Nonlinear Science and Numerical Simulation.* Vol 14, No 12, 4033-4040. (2009).
49. Y. S. Kivshar & B. A. Malomed. "Dynamics of solitons in nearly integrable systems". *Rev. Mod. Physics.* Vol 61, No 4, 763-915. (1989).
50. Y. Kodama & M. J. Ablowitz. "Perturbations of solitons and solitary waves". *Studies in Applied Mathematics.* Vol 64, 225-245. (1981).
51. R. Kohl, A. Biswas, D. Milovic & E. Zerrad. "Perturbation of Gaussian optical solitons in dispersion-managed fibers". *Applied Mathematics and Computation.* Vol 199, No 1, 250-258. (2009).
52. R. Kohl, A. Biswas, D. Milovic & E. Zerrad. "Perturbation of super-sech solitons in dispersion-managed optical fibers". *International Journal of Theoretical Physics.* Vol 47, No 7, 2038-2064. (2008).
53. R. Kohl, A. Biswas, D. Milovic & E. Zerrad. "Adiabatic dynamics of Gaussian and super-Gaussian solitons in dispersion-managed optical fibers". *Progress in Electromagnetics Research.* Vol 84, 27-53. (2008).
54. R. Kohl, D. Milovic, E. Zerrad & A. Biswas. "Perturbation of super-Gaussian optical solitons in dispersion-managed fibers". *Mathematical and Computer Modelling.* Vol 49, No 7-8, 418-427. (2009).

55. R. Kohl, D. Milovic, E. Zerrad & A. Biswas. "Optical solitons by He's variational principle in a non-Kerr law media". *Journal of Infrared, Millimeter and Terahertz Waves.* Vol 30, No 5, 526-537. (2009).
56. R. Kohl, D. Milovic, E. Zerrad & A. Biswas. "Soliton perturbation theory for dispersion-managed optical fibers". *Journal of Nonlinear Optical Physics and Materials.* Vol 18, No 2. (2009).
57. Z. M. Liao, C. J. McKinstrie & G. P. Agrawal. "Importance of pre-chirping in constant-dispersion fiber links with a large amplifier spacing". *Journal of Optical Society of America B.* Vol 17, No 4, 514-518. (2000).
58. P. M. Lushnikov. "Dispersion-managed solitons in optical fibers with zero average dispersion". *Optics Letters.* Vol 25, No 16, 1144-1146. (2000).
59. M. F. Mahmood & S. B. Qadri. "Modeling propagation of chirped solitons in an elliptically low birefringent single-mode optical fiber". *Journal of Nonlinear Optical Physics and Materials.* Vol 8, No 4, 469-475. (1999).
60. M. Matsumoto. "Analysis of interaction between stretched pulses propagating in dispersion-managed fibers". *IEEE Photonics Technology Letters.* Vol 10, No 3, 373-375. (1998).
61. C. R. Menyuk. "Application of multiple-length-scale methods to the study of optical fiber transmission". *Journal of Engineering Mathematics.* Vol 36, No 1-2, 113-136. (1999).
62. H. Michinel. "Pulses nonlinear surface waves and soliton emission at nonlinear graded index waveguides". *Optical and Quantum Electronics.* Vol 30, No 2, 79-97. (1998).
63. M. Nakazawa H. Kubota, K. Suzuki & E. Yamada. "Recent progress in soliton transmission technology". *Chaos.* Vol 10, No 3, 486-514. (2000).
64. T. Okamawari, A. Maruta & Y. Kodama. "Analysis of Gordon-Haus jitter in a dispersion-compensated optical transmission system". *Optics Letters.* Vol 23, No 9, 694-696. (1998).
65. T. Okamawari, A. Maruta & Y. Kodama. "Reduction of Gordon-Haus jitter in a dispersion-compensated optical transmission system: analysis". *Optics Communications.* Vol 141, No 9, 262-266. (1998).
66. A. Panajotovic, D. Milovic & A. Mittic. "Boundary case of pulse propagation analytic solution in the presence of interference and higher order dispersion". *TELSIKS 2005 Conference Proceedings.* 547-550. Nis-Serbia. (2005).
67. A. Panajotovic, D. Milovic, A. Biswas & E. Zerrad. "Influence of even order dispersion on super-sech soliton transmission quality under coherent crosstalk". *Research Letters in Optics.* Vol 2008, 613986, 5 pages. (2008).
68. A. Panajotovic, D. Milovic & A. Biswas. "Influence of even order dispersion on soliton transmission quality with coherent interference". *Progress in Electromagnetics Research B.* Vol 3, 63-72. (2008).
69. Y. Pointurier, M. Brandt-Pearce & C. L. Brown. "Analytical study of crosstalk propagation in all-optical networks using perturbation theory". *Journal of Lightwave Technology.* Vol 23, No 12, 4074-4083. (2005).
70. J. M. Senior. *Optic Fiber Communications.* Prentice Hall, New York, NY. USA. (1992).
71. V. N. Serkin & A. Hasegawa. "Soliton Management in the Nonlinear Schrodinger Equation Model with Varying Dispersion, Nonlinearity and Gain". *JETP Letters.* Vol 72, No 2, 89-92. (2000).
72. O. V. Sinkin, V. S. Grigoryan & C. R. Menyuk. "Accurate Probabilistic treatment of bit-pattern-dependent Nonlinear Distortions in BER calculations for WDM RZ systems". *Journal of Lightwave Technology.* Vol 25, No 10, 2959-2967. (2007).
73. C. D. Stacey, R. M. Jenkins, J. Banerji & A. R. Davis. "Demonstration of fundamental mode only propagation in highly multimode fibre for high power EDFAs". *Optics Communications.* Vol 269, 310-314. (2007).
74. M. Stefanovic & D. Milovic. "The impact of out-of-band crosstalk on optical communication link preferences". *Journal of Optical Communications.* Vol 26, No 2, 69-72. (2005).

75. M. Stefanovic, D. Draca, A. Panajotovic & D. Milovic. "Individual and joint influence of second and third order dispersion on transmission quality in the presence of coherent interference". *Optik.* Vol 120, No 13, 636-641.(2009).
76. B. Stojanovic, D. M. Milovic & A. Biswas. "Timing shift of optical pulses due to inter-channel crosstalk". *Progress in Electromagnetics Research M.* Vol 1, 21-30. (2008).
77. S. K. Turitsyn, I. Gabitov, E. W. Laedke, V. K. Mezentsev, S. L. Musher, E. G. Shapiro, T. Schäfer, & K. H. Spatschek. "Variational approach to optical pulse propagation in dispersion compensated transmission system". *Optics Communications.* Vol 151, No 1-3, 117-135. (1998).
78. S. Wabnitz, Y. Kodama. & A. B. Aceves "Control of optical soliton interactions". *Optical Fiber Technology.* Vol 1, 187-217. (1995).
79. P. K. A. Wai & C. R. Menyuk. "Polarization mode dispersion, de-correlation, and diffusion in optical fibers with randomly varying birefringence". *Journal of Lightwave Technology.* Vol 14, No 2, 148-157. (1996).
80. V. E. Zakharov & S. Wabnitz. *Optical Solitons: Theoretical Challenges and Industrial Perspectives.* Springer, New York, CA. USA. (1999).

Chapter 3
Polarization Preserving Fibers

3.1 Introduction

Birefringence in conventional single-mode fibers changes randomly due to variations in the core shape and anisotropic stress acting on the core. Linearly polarized light launched into the fiber reaches very quickly into a state of arbitrary polarization. Pulse broadening is a result of the fact that different frequency components of an optical pulse acquire different polarization states. This is known as *polarization mode dispersion* (PMD) and is a limiting factor for terrestrial high bit rate optical communication systems. Nowadays it is possible to manufacture fibers for which random fluctuations in the core shape and size are not the governing factor in determining the state of polarization. Such fibers are called *polarization preserving* (PP) fibers. A large amount of birefringence is introduced intentionally in these fibers by the aid of design modifications so that small random birefringence fluctuations do not affect the light polarization significantly.

Polarization preserving optical fibers preserves the plane of polarization of the light launched into it. This type of fiber is a single-mode fiber and is also called the *polarization maintaining* (PM) *fiber*. The polarization is preserved by introducing asymmetry in the fiber structure. This asymmetry may be either in fiber internal stresses (stress-induced birefringence) or in fiber shape (geometrical birefringence). Asymmetry causes different propagation constants for two perpendicular polarized modes that are transmitted by the fiber. Cross-coupling between these modes are reduced as compared to the conventional single-mode fiber.

Polarization multiplexing should not work unless polarization-preserving fibers are not used. It turns out that even though polarization states of the bit train does change in an unpredictable manner, the orthogonal nature of any two neighboring bits is nearly preserved. Because of this orthogonality, the interaction among solitons is much weaker as compared to the co-polarized solitons case. The reduced interaction lowers the bit-error rate (BER) and increases the transmission distance of a Gb/s soliton system.

The PP fiber has a higher attenuation than conventional single-mode fiber. These are the main reasons why this type of fiber is rarely used for long-distance transmission. They are instead commonly used for telecommunication applications, fiber-optic sensing and interferometry. Besides that PM fibers are not optically symmetrical and have strong internal birefringence caused by stress applying members. The internal birefringence is significantly higher than normal levels of bend-induced birefringence. In this way, it preserves the state of polarization when the laser beam is correctly aligned to either of the two axes.

The two most critical measures of a terminated PP fiber are extinction ratio and key alignment accuracy. Extinction ratio can be easily degraded by inappropriate adhesives or untested curing procedures. A PM fiber can maintain a 25–35 dB extinction ratio but after concatenation, this figure can be as low as 10–15 dB if not processed with great care. The fiber orientation misalignment can cause a crosstalk between TE and TM modes even when the extinction ratio is good. For example, a perfectly terminated PM fiber with 27 dB extinction ratio can generate 15 dB crosstalk due to key misalignment. The fiber length is also a potential source of extinction ratio degradation and it is extremely important to keep the entire fiber free from twists, bends and temperature variations.

The dimensionless form of DM-NLSE, in polarization preserving fibers, is given by

$$iu_z + \frac{D(z)}{2}u_{tt} + g(z)|u|^2u = 0 \tag{3.1}$$

This equation governs the propagation of optical solitons in 1+1 dimensions. It is derived in the previous chapter. The mathematical structure of the optical pulse for (3.1) is taken to be

$$\begin{aligned} u(z,t) &= A(z)f\left[B(z)\left\{t-\bar{t}(z)\right\}\right] \\ &\quad \cdot \exp[iC(z)\left\{t-\bar{t}(z)\right\}^2 - i\kappa(z)\left\{t-\bar{t}(z)\right\} + i\theta(z)] \end{aligned} \tag{3.2}$$

where f represents the functional form of the pulse while the interpretation of other soliton parameters are already discussed in the previous introductory chapter.

3.2 Integrals of motion

Recall that the two integrals of motion of the polarization preserving fibers are given by [25]

$$E = \int_{-\infty}^{\infty} |u|^2 dt = \frac{A^2}{B} I_{0,2,0} \tag{3.3}$$

and

$$M = \frac{i}{2} D(z) \int_{-\infty}^{\infty} (u^* u_t - u u_t^*) dt = -\kappa D(z) \frac{A^2}{B} I_{0,2,0} \tag{3.4}$$

These are respectively the *energy* and the *linear momentum* of the soliton. Also, it was mentioned in the previous chapter that the third important quantity, namely the Hamiltonian, is given by

$$H = \frac{1}{2} \int_{-\infty}^{\infty} \left(D(z)|u_t|^2 - g(z)|u|^4 \right) dt$$

$$= \frac{D(z)}{2} \left(A^2 B I_{0,0,2} + 4\frac{A^2 C^2}{B^3} I_{2,2,0} + \frac{\kappa^2 A^2}{B} I_{0,2,0} \right) - \frac{g(z)}{2} \frac{A^4}{B} I_{0,4,0} \tag{3.5}$$

is not an integral of the motion, unless $D(z)$ and $g(z)$ are constants, in which there actually exists infinitely many conserved quantities.

In the following three subsections, these conserved quantities and the Hamiltonian of solitons are going to be computed for the different kind of pulses that are going to be studied in this book. Although there are various kinds of pulses that are studied in the literature of DM solitons, the Gaussian, super-Gaussian and the super-sech type pulses are only considered in this book. The other kinds of pulses that are not touched upon in this text are Gauss-Hermite pulses [40], cosh-Gaussian pulses and sinh-Gaussian pulses [25] and many more.

3.2.1 Gaussian pulses

For a pulse of Gaussian type, $f(\tau) = e^{-\frac{\tau^2}{2}}$. So, the pulse format is given by

$$u(z,t) = A(z) e^{-\frac{1}{2}B^2(z)\left(t-\bar{t}(z)\right)^2}$$

$$\cdot \exp[iC(z)\left\{t-\bar{t}(z)\right\}^2 - i\kappa(z)\left\{t-\bar{t}(z)\right\} + i\theta(z)] \tag{3.6}$$

Figure.3.1 shows the plot of a Gaussian profile.

Thus, the conserved quantities respectively reduce to

$$E = \int_{-\infty}^{\infty} |u|^2 dt = \frac{A^2}{B} \sqrt{\pi} \tag{3.7}$$

$$M = \frac{i}{2} D(z) \int_{-\infty}^{\infty} (u^* u_t - u u_t^*)\, dt = -\kappa D(z) \frac{A^2}{B} \sqrt{\pi} \tag{3.8}$$

while the Hamiltonian is

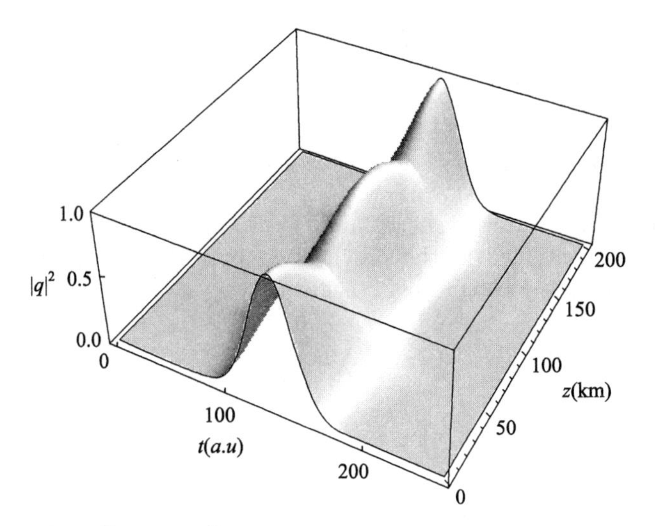

Fig. 3.1 Breathing Gaussian soliton.

$$\begin{aligned} H &= \frac{1}{2}\int_{-\infty}^{\infty} [D(z)|u_t|^2 - g(z)|u|^4]dt \\ &= \frac{\sqrt{\pi}A^2}{4B^3}\left\{D(z)\left(B^4 - 2\kappa^2 B^2 + 4C^2\right) - \sqrt{2}g(z)A^2B^2\right\} \end{aligned} \tag{3.9}$$

Figure 3.2 shows the root-mean square variation of the width of the Gaussian soliton.

3.2.2 Super-Gaussian pulses

For a super-Gaussian (SG) pulse, $f(\tau) = e^{-\frac{\tau^{2m}}{2}}$ with $m \geq 1$ where the parameter m controls the degree of edge sharpness. With $m = 1$, the case of a chirped Gaussian pulse is recovered while for larger values of m the pulse gradually becomes square shaped with sharper leading and trailing edges [25]. The pulse profile is given by

$$\begin{aligned} u(z,t) &= A(z)e^{-\frac{1}{2}B^{2m}(z)\left(t-\bar{t}(z)\right)^{2m}} \\ &\quad \cdot \exp[iC(z)\left\{t - \bar{t}(z)\right\}^2 - i\kappa(z)\left\{t - \bar{t}(z)\right\} + i\theta(z)] \end{aligned} \tag{3.10}$$

In Figure 3.3, one can see the shapes of the pulses as the parameter m varies.

For an SG pulse, the integrals of motion respectively are

$$E = \int_{-\infty}^{\infty} |u|^2 dt = \frac{A^2}{mB}\Gamma\left(\frac{1}{2m}\right) \tag{3.11}$$

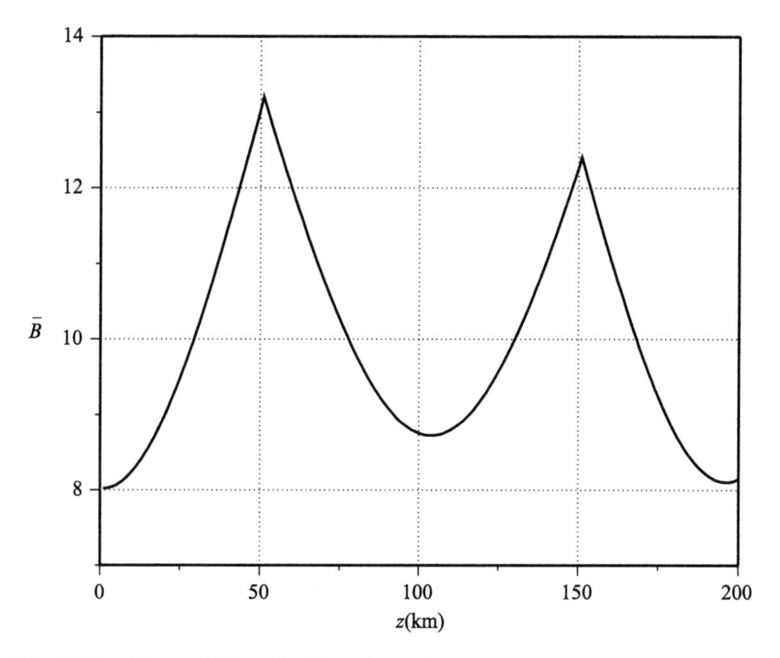

Fig. 3.2 RMS width variation of a Gaussian pulse.

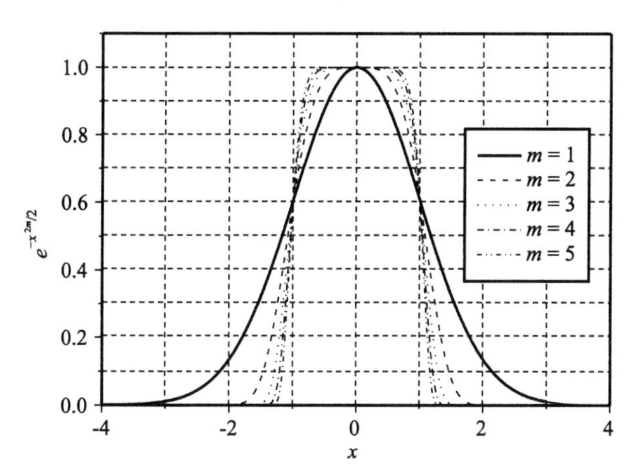

Fig. 3.3 SG pulses for various values of m.

$$M = \frac{i}{2}D(z)\int_{-\infty}^{\infty}(u^* u_t - u u_t^*)\,dt = -\kappa D(z)\frac{A^2}{mB}\Gamma\left(\frac{1}{2m}\right) \tag{3.12}$$

while the Hamiltonian is

$$H = \frac{1}{2}\int_{-\infty}^{\infty}\left[D(z)|u_t|^2 - g(z)|u|^4\right]dt$$

$$= \frac{A^2}{B}\left[D(z)\left\{\frac{2C^2}{mB^2}\left(\frac{3}{2m}\right)+\frac{mB^2}{2}\Gamma\left(\frac{3}{2m}\right)-\frac{\kappa^2}{2m}\Gamma\left(\frac{1}{2m}\right)\right\}\right.$$
$$\left. - g(z)\frac{A^2}{2m2^{\frac{1}{2m}}}\Gamma\left(\frac{1}{2m}\right)\right] \quad (3.13)$$

Figures 3.4 and 3.5 display the breathing SG pulse and the root-mean square variation of the soliton.

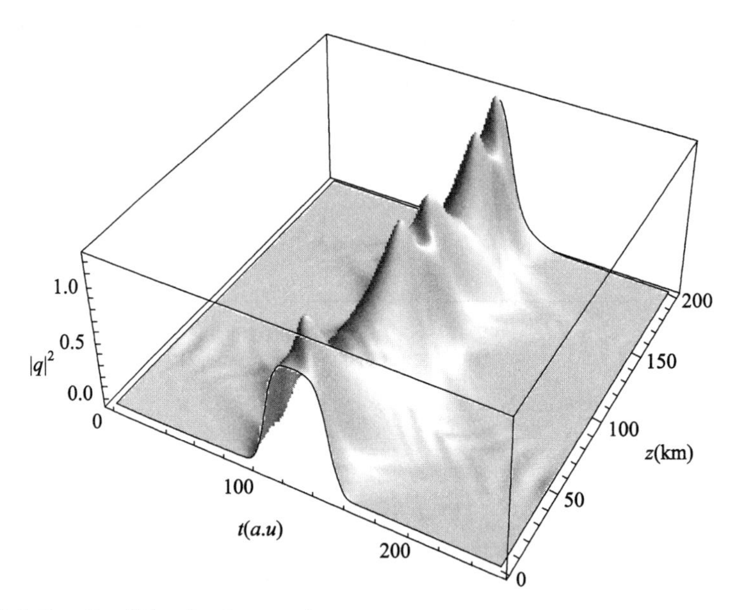

Fig. 3.4 Breating SG pulse for $m = 2$.

3.2.3 Super-Sech pulses

Now, it is assumed that the solution of Eq.(3.1) is given by a super-sech (SS) chirped pulse $f(\tau) = 1/\cosh^n \tau$. Thus, the pulse profile is given by

$$u(z,t) = \frac{A(z)}{\cosh^n\left[B(z)\left(t-\bar{t}\right)\right]}$$
$$\cdot \exp[iC(z)\left\{t-\bar{t}(z)\right\}^2 - i\kappa(z)\left\{t-\bar{t}(z)\right\} + i\theta(z)] \quad (3.14)$$

Here, the parameter $n > 0$. The SS pulses for various values of the parameter n is given in Figure 3.6.

For SS pulses, the integrals of motion are given as

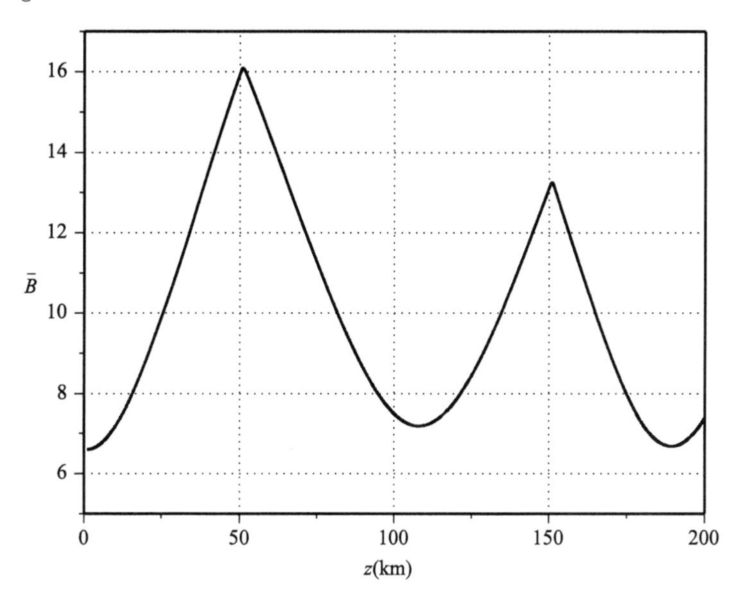

Fig. 3.5 RMS width variation of an SG pulse for $m = 2$.

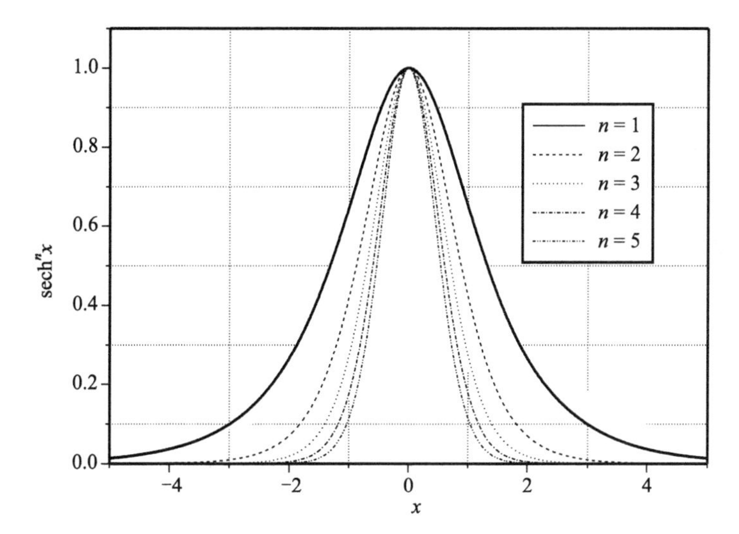

Fig. 3.6 SS pulses for various values of n.

$$E = \int_{-\infty}^{\infty} |u|^2 dt = \frac{A^2}{B} B\left(n, \frac{1}{2}\right) \tag{3.15}$$

$$M = \frac{i}{2} D(z) \int_{-\infty}^{\infty} \left(u^* u_t - u u_t^*\right) dt = -\kappa D(z) \frac{A^2}{B} B\left(n, \frac{1}{2}\right) \tag{3.16}$$

while the Hamiltonian is

$$
\begin{aligned}
H &= \frac{1}{2}\int_{-\infty}^{\infty}\left[D(z)|u_t|^2 - g(z)|u|^4\right]dt \\
&= \frac{A^2}{B}\left\{\frac{D(z)}{2}\left[\left(\frac{n^2B^2}{2n+1}-\kappa^2\right)B\left(n,\frac{1}{2}\right)\right.\right. \\
&\quad \left. + \frac{2^{2n}C^2}{B^2n^3}\,{}_4F_3\left(n,n,n,2n;n+1,n+1,n+1;-1\right)\right] \\
&\quad \left. - \frac{g(z)A^2}{2}B\left(2n,\frac{1}{2}\right)\right\}
\end{aligned} \tag{3.17}
$$

where the Gauss' generalized hyper-geometric function is defined as

$$
{}_pF_q\left(a_1,\ldots,a_p;b_1,\ldots,b_q;x\right) = \sum_{k=0}^{n}\frac{\prod_{j=1}^{p}\left(a_j\right)_k x^k}{\prod_{j=1}^{q}\left(b_j\right)_k k!} \tag{3.18}
$$

where $B(l,m)$ is the beta function and the Poehammer symbol or the rising factorial notation $(m)_j$ is defined as

$$
\begin{aligned}
(m)_n &= m(m+1)(m+2)\cdots(m+n-1) \\
&= \frac{(m+n-1)!}{(m-1)!} = \frac{\Gamma(m+n)}{\Gamma(m)}
\end{aligned} \tag{3.19}
$$

Figures 3.7 and 3.8 show the profile of a super-sech pulse and the root-mean square variation of the width of an SS pulse.

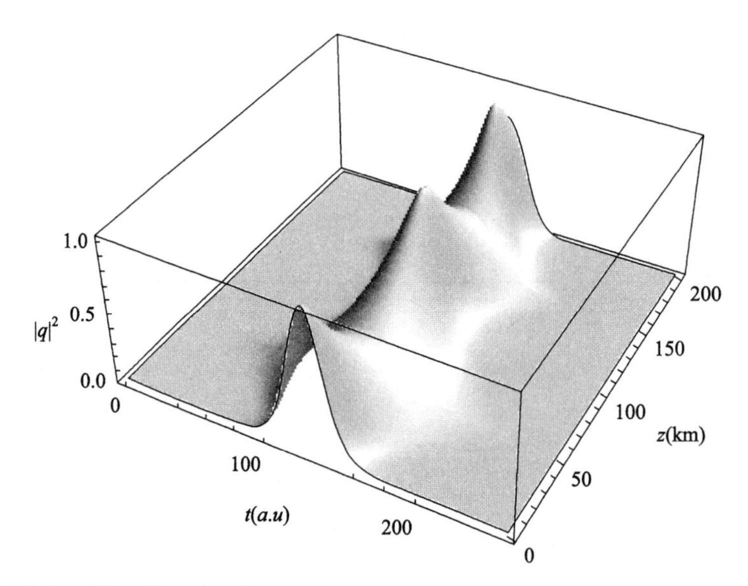

Fig. 3.7 Breathing SS pulses for $n = 2$.

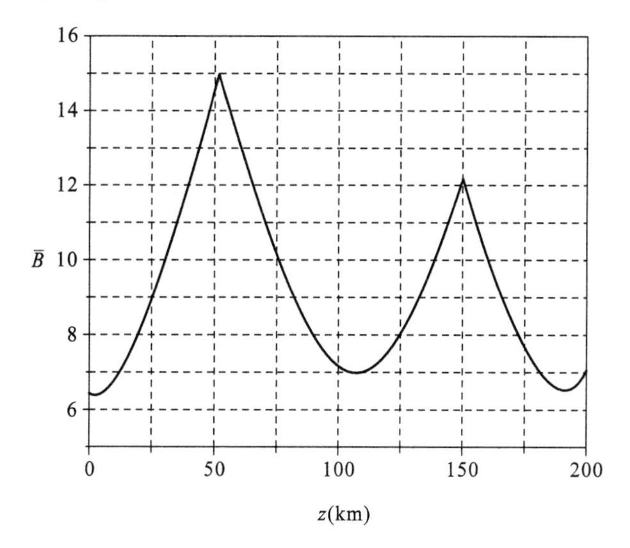

Fig. 3.8 RMS width variation for SS pulse for $n = 2$.

3.3 Variational principle

The variational principle that was derived in the previous chapter will now be used to obtain the soliton parameter dynamics for Gaussian and super-Gaussian pulses. It needs to be noted that it is now left to the reader, as an exercise, to obtain these parameter dynamics for super-sech pulses, in a very similar fashion. The Lagrangian given by Eq.(2.84) is computed for the Gaussian and super-Gaussian pulses and then substituting the six soliton parameters for p is Eq.(2.85) yields the following parameter dynamics in the next two subsections.

3.3.1 Gaussian pulses

$$\frac{dA}{dz} = -ACD(z) \tag{3.20}$$

$$\frac{dB}{dz} = -2BCD(z) \tag{3.21}$$

$$\frac{dC}{dz} = 2D(z)\left(B^4 - C^2\right) - \frac{\sqrt{2}}{2}g(z)A^4B \tag{3.22}$$

$$\frac{d\kappa}{dz} = 0 \tag{3.23}$$

$$\frac{d\bar{t}}{dz} = -\kappa D(z) \tag{3.24}$$

$$\frac{d\theta}{dz} = \frac{D(z)}{2}\left(\kappa^2 - 2B^2\right) + \frac{5\sqrt{2}}{8} g(z) A^2 \tag{3.25}$$

Equations (3.20)–(3.25) represent the evolution equations of the parameters of a Gaussian soliton propagating through an optical fiber. These evolution equations can be used to study various issues including the pulse interaction.

3.3.2 Super-Gaussian pulses

$$\frac{dA}{dz} = -ACD(z) \tag{3.26}$$

$$\frac{dB}{dz} = -2BCD(z) \tag{3.27}$$

$$\frac{dC}{dz} = \left(B^4 \frac{m^2}{2^{\frac{m-2}{m}}} \frac{\Gamma\left(\frac{4m-1}{2m}\right)}{\Gamma\left(\frac{3}{2m}\right)} - 2C^2\right) D(z) - g(z) A^4 B \frac{1}{2^{\frac{4m+1}{2m}}} \frac{\Gamma\left(\frac{1}{2m}\right)}{\Gamma\left(\frac{3}{2m}\right)} \tag{3.28}$$

$$\frac{d\kappa}{dz} = 0 \tag{3.29}$$

$$\frac{d\bar{t}}{dz} = -\kappa D(z) \tag{3.30}$$

$$\frac{d\theta}{dz} = \left(\frac{\kappa^2}{2} - m^2 2^{\frac{1}{m}} \frac{\Gamma\left(\frac{4m-1}{2m}\right)}{\Gamma\left(\frac{1}{2m}\right)} B^2\right) D(z) + 5g(z) A^2 \frac{1}{2^{\frac{4m+1}{2m}}} \tag{3.31}$$

Note, here, that for $m = 1$, Eqs.(3.26)–(3.31) all reduce to Eqs.(3.20)–(3.25) respectively for Gaussian solitons.

3.4 Perturbation terms

In this section, similarly the adiabatic parameter dynamics of Gaussian and SG solitons will be obtained. This is also based on the variational principle. The starting point is the EL equation given by Eq.(2.93). Subsequently, relations (2.94)–(2.99) respectively reduce to the following set of relations of the soliton parameters that are described in the following two subsections for Gaussian and SG pulses, respectively.

3.4.1 Gaussian pulses

$$\frac{dA}{dz} = -ACD(z) - \frac{\epsilon}{\sqrt{2\pi}} \int_{-\infty}^{\infty} \Re[Re^{-i\phi}](4\tau^2 - 3)e^{-\tau^2} d\tau \tag{3.32}$$

$$\frac{dB}{dz} = -2BCD(z) - \epsilon\sqrt{\frac{2}{\pi}\frac{B}{A}} \int_{-\infty}^{\infty} \Re[Re^{-i\phi}](4\tau^2 - 1)e^{-\tau^2} d\tau \tag{3.33}$$

$$\begin{aligned} \frac{dC}{dz} &= 2D(z)\left(B^4 - C^2\right) - \frac{g(z)A^2B^2}{\sqrt{2}} \\ &\quad - 2\epsilon\sqrt{\frac{2}{\pi}\frac{B^2}{A}} \int_{-\infty}^{\infty} \Im[Re^{-i\phi}](1 - 4\tau^2)e^{-\tau^2} d\tau \end{aligned} \tag{3.34}$$

$$\frac{d\kappa}{dz} = -\epsilon\sqrt{\frac{2}{\pi}\frac{4}{AB}} \int_{-\infty}^{\infty} \left\{\tau B^2 \Im[Re^{-i\phi}] + \tau C \Re[Re^{-i\phi}]\right\} e^{-\tau^2} d\tau \tag{3.35}$$

$$\frac{d\bar{t}}{dz} = -\kappa D(z) + \epsilon\frac{\sqrt{2\pi}}{AB} \int_{-\infty}^{\infty} \Re[Re^{-i\phi}]\tau e^{-\tau^2} d\tau \tag{3.36}$$

$$\begin{aligned} \frac{d\theta}{dz} &= \frac{D(z)}{2}(\kappa^2 - 2B^2) + \frac{5\sqrt{2}}{8}g(z)A^2 \\ &\quad + \frac{\epsilon}{\sqrt{2\pi}}\frac{1}{AB} \int_{-\infty}^{\infty} \{B\Im[Re^{-i\phi}](3 - 4\tau^2) \\ &\quad + 4\tau\kappa\Re[Re^{-i\phi}]\tau\} e^{-\tau^2} d\tau \end{aligned} \tag{3.37}$$

These equations now represent the evolution equations for the parameters of a Gaussian pulse propagating through an optical fiber in presence of the perturbation terms.

3.4.2 Super-Gaussian pulses

$$\begin{aligned} \frac{dA}{dz} &= -ACD(z) - \epsilon\frac{m^2 2^{\frac{m+2}{m}}}{\Gamma\left(\frac{1}{2m}\right)\Gamma\left(\frac{3}{2m}\right)} \int_{-\infty}^{\infty} \Re[Re^{-i\phi}] \\ &\quad \cdot \left\{\frac{\tau^2}{m2^{\frac{1}{m}}}\Gamma\left(\frac{1}{2m}\right) - \frac{3}{m2^{\frac{3}{2m}}}\Gamma\left(\frac{3}{2m}\right)\right\} e^{-\tau^{2m}} d\tau \end{aligned} \tag{3.38}$$

$$\frac{dB}{dz} = -2BCD(z) - \epsilon\frac{B}{A}\frac{m^2 2^{\frac{2}{m}}}{\Gamma\left(\frac{1}{2m}\right)\Gamma\left(\frac{3}{2m}\right)} \int_{-\infty}^{\infty} \Re[Re^{-i\phi}]$$

$$\cdot \left\{ \frac{\tau^2}{m2^{\frac{1}{2m}}} \Gamma\left(\frac{1}{2m}\right) - \frac{1}{m2^{\frac{3}{2m}}} \Gamma\left(\frac{3}{2m}\right) \right\} e^{-\tau^{2m}} d\tau \tag{3.39}$$

$$\begin{aligned} \frac{dC}{dz} = & \left(B^4 \frac{m^2}{2^{\frac{m-2}{m}}} \frac{\Gamma\left(\frac{4m-1}{2m}\right)}{\Gamma\left(\frac{3}{2m}\right)} - 2C^2 \right) D(z) \\ & - g(z) A^2 B^2 \frac{1}{2^{\frac{4m+1}{2m}}} \frac{\Gamma\left(\frac{1}{2m}\right)}{\Gamma\left(\frac{3}{2m}\right)} \\ & - \epsilon \frac{B^2}{A} \frac{m2^{\frac{2m+3}{2m}}}{\Gamma\left(\frac{3}{2m}\right)} \int_{-\infty}^{\infty} \Im[Re^{-i\phi}](1 - 4m\tau^{2m}) e^{-\tau^{2m}} d\tau \end{aligned} \tag{3.40}$$

$$\begin{aligned} \frac{d\kappa}{dz} = & - \epsilon \frac{1}{AB} \frac{m2^{\frac{2m-1}{2m}}}{\Gamma\left(\frac{1}{2m}\right)} \int_{-\infty}^{\infty} \{2m\tau^{2m-1} B^2 \Im[Re^{-i\phi}] \\ & + 2\tau C \Re[Re^{-i\phi}]\} e^{-\tau^{2m}} d\tau \end{aligned} \tag{3.41}$$

$$\frac{d\bar{t}}{dz} = -\kappa D(z) + \epsilon \frac{1}{AB} \frac{m2^{\frac{2m+1}{2m}}}{\Gamma\left(\frac{1}{2m}\right)} \int_{-\infty}^{\infty} \Re[Re^{-i\phi}] \tau e^{-\tau^{2m}} d\tau \tag{3.42}$$

$$\begin{aligned} \frac{d\theta}{dz} = & \left(\frac{\kappa^2}{2} - m^2 2^{\frac{1}{m}} \frac{\Gamma\left(\frac{4m-1}{2m}\right)}{\Gamma\left(\frac{1}{2m}\right)} B^2 \right) D(z) + 5g(z) A^2 \frac{1}{2^{\frac{4m+1}{2m}}} \\ & + \epsilon \frac{1}{AB} \frac{m2^{\frac{2m+1}{2m}}}{\Gamma\left(\frac{1}{2m}\right)} \int_{-\infty}^{\infty} \{B \Im[Re^{-i\phi}](3 - 4m\tau^{2m}) \\ & + 4\kappa \Re[Re^{-i\phi}] \tau\} e^{-\tau^{2m}} d\tau \end{aligned} \tag{3.43}$$

So, now, these are the adiabatic evolution of the soliton parameters for an SG pulse in presence of the perturbation terms.

3.5 Stochastic perturbation

Besides these issues, one also needs to take into account, from practical considerations, the stochastic aspects. These effects can be classified into three basic types [1, 14]:

1. Stochasticity associated with the chaotic nature of the initial pulse due to partial coherence of the laser generated radiation.
2. Stochasticity due to random non-uniformities in the optical fibers like the fluctuations in the values of dielectric constant the random variations of the fiber diameter and more.

3. The chaotic field caused by a dynamic stochasticity might arise from a periodic modulation of the system parameters or when a periodic array of pulses propagate in a fiber optic resonator.

Thus, stochasticity is inevitable in optical soliton communications.

Stochasticity are basically of two types, namely, homogeneous and non-homogeneous [1, 14]. In inhomogeneous case, the stochasticity is present in the input pulse of the fiber. So, the parameter dynamics are deterministic but however the initial values are random. While in the homogeneous case the stochasticity originates due to the random perturbation of the fiber like the density fluctuation of the fiber material or the random variations in the fiber diameter etc.

The perturbed NLSE given by Eq.(2.75) is going to be studied in this section, where the perturbation term R will now include a random perturbation term. Thus, in this section,

$$R = \alpha u + \beta u_{tt} + \sigma(z,t) \tag{3.44}$$

Here in (3.44), $\alpha < 0$ (> 0) is the attenuation (amplification) coefficient, β is the coefficient of bandpass filtering. Also, $\sigma(z,t)$ represents the random stochastic perturbation term.

The use of optical amplifiers affects the evolution of solitons considerably. The reason is that amplifiers, although needed to restore soliton energy, introduces noise originating from amplified spontaneous emission (ASE) [14]. To study the impact of noise on soliton evolution, the evolution of the mean energy of the soliton due to ASE will be studied in this section. In case of lumped amplification, solitons are perturbed by ASE in a discrete fashion at the location of the amplifiers. It can be assumed that noise is distributed all along the fiber length since the amplifier spacing satisfies $z_a \ll 1$. In (3.44), $\sigma(z,t)$ represents the noise term with Gaussian statistics and is assumed that $\sigma(z,t)$ [14] is a function of z only so that $\sigma(z,t) = \sigma(z)$. Now, the complex stochastic term $\sigma(z)$ can be decomposed into real and imaginary parts as

$$\sigma(z) = \sigma_1(z) + i\sigma_2(z) \tag{3.45}$$

It is further assumed that $\sigma_1(z)$ and $\sigma_2(z)$ are independently delta correlated:

$$\langle \sigma_1(z) \rangle = \langle \sigma_2(z) \rangle = \langle \sigma_1(z)\sigma_2(z') \rangle = 0 \tag{3.46}$$

$$\langle \sigma_1(z)\sigma_1(z') \rangle = 2D_1\delta(z-z') \tag{3.47}$$

$$\langle \sigma_2(z)\sigma_2(z') \rangle = 2D_2\delta(z-z') \tag{3.48}$$

where D_1 and D_2 are related to the ASE spectral density. In this case, it is assumed that $D_1 = D_2 = \overline{D}$. Thus,

$$\langle \sigma(z) \rangle = 0 \tag{3.49}$$

and

$$\langle \sigma(z)\sigma(z')\rangle = 2\delta(z-z') \tag{3.50}$$

In soliton units, one gets

$$\overline{D} = \frac{F_n F_G}{N_{ph} z_a} \tag{3.51}$$

where F_n is the amplifier noise figure, while

$$F_G = \frac{(G-1)^2}{G \ln G} \tag{3.52}$$

is related to the amplifier gain G and finally N_{ph} is the average number of photons in the pulse propagating as a fundamental soliton [14].

In presence of the perturbation terms, given by Eq.(3.44), the adiabatic dynamics of the soliton energy (E) and the frequency (κ) are given by

$$\begin{aligned}\frac{dE}{dz} = &\frac{\epsilon}{mB^3}\left[2\alpha A^2 B^2 \Gamma\left(\frac{1}{2m}\right) - 8\beta A^2 C^2 \Gamma\left(\frac{3}{2m}\right)\right.\\ &\left. - 2\beta\kappa^2 A^2 B^2 \Gamma\left(\frac{1}{2m}\right) - \beta A^2 B^4 m(2m-1)\Gamma\left(\frac{2m-1}{2m}\right)\right]\\ &+ \frac{2\epsilon A}{B}\int_{-\infty}^{\infty} e^{-\frac{1}{2}\tau^{2m}}\left(\sigma_1\cos\phi + \sigma_2 \sin\phi\right) d\tau\end{aligned} \tag{3.53}$$

$$\begin{aligned}\frac{d\kappa}{dz} = &-\frac{4\epsilon\beta\kappa A^2}{mB^3 Ed(z)}\left[\kappa^2 A^2 B^2 \Gamma\left(\frac{1}{2m}\right) + 8C^2\Gamma\left(\frac{3}{2m}\right)\right]\\ &+ \frac{\epsilon}{B^2 Ed(z)}\int_{-\infty}^{\infty} e^{-\frac{1}{2}\tau^{2m}}\left[4AC\tau\left(\sigma_1\cos\phi + \sigma_2\sin\phi\right)\right.\\ &\left.+ 2mAB^2\tau^{2m-1}\left(\sigma_2\cos\phi - \sigma_1\sin\phi\right)\right] d\tau\end{aligned} \tag{3.54}$$

which are obtained by the aid of Eqs.(2.76) and (2.77), respectively. Here in Eqs.(3.53) and (3.54), τ and ϕ are as in Eqs.(2.100) and (2.101), respectively. Equations (3.53) and (3.54), as it appears, are difficult to analyze. If the terms with σ_1 and σ_2 are suppressed, one recovers a dynamical system, for a fixed chirp, which has a stable fixed point, namely a sink, of Eqs.(3.53) and (3.54) given by $(\overline{B}, \overline{\kappa})$ where

$$\overline{B} = \left[\frac{\alpha\Gamma\left(\frac{1}{2m}\right) + \sqrt{\alpha^2\Gamma\left(\frac{1}{2m}\right) - 8\beta^2 C^2 m(2m-1)\Gamma\left(\frac{3}{2m}\right)\Gamma\left(\frac{2m-1}{2m}\right)}}{\beta m(2m-1)\Gamma\left(\frac{2m-1}{2m}\right)}\right]^{1/2} \tag{3.55}$$

$$\overline{\kappa} = 0 \tag{3.56}$$

Now, linearizing about this fixed point, Eqs.(3.53) and (3.54) respectively reduce to

$$\frac{dE}{dz} = \epsilon\left[2\alpha E - \beta\overline{B}^2 m(2m-1)\frac{\Gamma\left(\frac{2m-1}{2m}\right)}{\Gamma\left(\frac{3}{2m}\right)} - \frac{8\beta C^2 E}{\overline{B}^2}\frac{\Gamma\left(\frac{3}{2m}\right)}{\Gamma\left(\frac{1}{2m}\right)}\right]$$

$$+ \frac{2\epsilon m E}{A\Gamma\left(\frac{1}{2m}\right)}\int_{-\infty}^{\infty} e^{-\frac{1}{2}\tau^{2m}}(\sigma_1\cos\phi + \sigma_2\sin\phi)d\tau \tag{3.57}$$

$$\frac{d\kappa}{dz} = -\frac{32\epsilon\kappa\beta C^2}{\overline{B}^2 D(z)}\frac{\Gamma\left(\frac{3}{2m}\right)}{\Gamma\left(\frac{1}{2m}\right)}$$

$$+\frac{4\epsilon m C}{ABd(z)\Gamma\left(\frac{1}{2m}\right)}\int_{-\infty}^{\infty}\tau e^{-\frac{1}{2}\tau^{2m}}(\sigma_1\cos\phi + \sigma_2\sin\phi)d\tau$$

$$+\frac{2m^2\epsilon B}{Ad(z)\Gamma\left(\frac{1}{2m}\right)}\int_{-\infty}^{\infty}\tau^{2m-1}e^{-\frac{1}{2}\tau^{2m}}(\sigma_2\cos\phi - \sigma_1\sin\phi)d\tau \tag{3.58}$$

Equations (3.57) and (3.58) are known as *Langevin equations* which will now be analyzed further to obtain the long term behavior of the SG pulses in presence of the stochastic perturbations.

After rescaling, Eq.(3.57) reduces to

$$\frac{dE}{dz} = \epsilon\left[E - \zeta(1-E)\right] \tag{3.59}$$

where

$$\zeta = \int_{-\infty}^{\infty} e^{-\frac{1}{2}\tau^{2m}}\left(\sigma_1\cos\phi + \sigma_2\sin\phi\right)d\tau \tag{3.60}$$

The stochastic phase factor is, now, defined by

$$\phi(z,y) = \int_y^z \zeta(s)ds \tag{3.61}$$

where $z > y$. Assuming that ζ is a Gaussian stochastic variable one gets

$$\left\langle e^{\phi(z,y)}\right\rangle = e^{\delta(z-y)} \tag{3.62}$$

$$\left\langle e^{[\phi(z,y)+\phi(z',y')]}\right\rangle = e^{\delta\theta} \tag{3.63}$$

where

$$\theta = 2(z+z'-y-y') - |z-z'| - |y-y'| \tag{3.64}$$

and

$$\langle\zeta(y)e^{-\phi(z,y)}\rangle = \frac{\partial}{\partial y}\langle e^{-\phi(z,y)}\rangle = \delta e^{\overline{D}(z-y)} \tag{3.65}$$

$$\langle\zeta(y)\zeta(y')e^{[-\phi(z,y)-\phi(z',y')]}\rangle = 2\overline{D}\delta(y-y')e^{\overline{D}\theta} + \frac{\partial^2}{\partial y\partial y'}e^{\delta\theta} \tag{3.66}$$

Now, solving Eq.(3.59) gives

$$\langle E(z)\rangle = \frac{\overline{D}E_0}{1-\overline{D}}\left\{1-e^{-\epsilon z(1-\overline{D})}\right\} \tag{3.67}$$

where, the initial energy E_0 is given by

$$E_0 = E(0) = \frac{mA^2}{B}\Gamma\left(\frac{1}{2m}\right) \tag{3.68}$$

From Eq.(3.67), it follows that

$$\lim_{z\to\infty}\langle E(z)\rangle = \frac{\overline{D}E_0}{1-\overline{D}} \tag{3.69}$$

Thus, it shows that in presence of the stochastic perturbation term of the form that is considered in this section, the SG pulses will propagate down the fiber with a fixed mean energy that is given by (3.69) as long as $\overline{D} < 1$. However, if $\overline{D} > 1$, $\langle E(z)\rangle$ becomes unbounded for large z.

References

1. F. Abdullaev, S. Darmanyan & P. Khabibullaev. *Optical Solitons.* Springer, New York, NY. USA. (1993).
2. M. J. Ablowitz & H. Segur. *Solitons and the Inverse Scattering Transform.* SIAM. Philadelphia, PA. USA. (1981).
3. M. J. Ablowitz & G. Biondini. "Multiscale pulse dynamics in communication systems with strong dispersion management". *Optics Letters.* Vol 23, No 21, 1668-1670. (1998).
4. M. J. Ablowitz & P. A. Clarkson. "Solitons and symmetries". *Journal of Engineering Mathematics.* Vol 36, 1-9. (1999).
5. M. J. Ablowitz & T. Hirooka. "Nonlinear effects in quasi-linear dispersion-managed systems". *IEEE Photonics Technology Letters.* Vol 13, No 10, 1082-1084. (2001).
6. G. P. Agrawal. *Nonlinear Fiber Optics.* Academic Press, San Deigo, CA. USA. (1995).
7. N. N. Akhmediev & A. Ankiewicz. *Solitons, Nonlinear Pulses and Beams.* Chapman and Hall, London. UK. (1997).
8. D. Anderson. "Variational approach to nonlinear pulse propagation in optical fibers". *Physical Review A.* Vol 27, No 6, 3135-3145. (1983).
9. A. Biswas. "Super-Gaussian solitons in optical fibers". *Fiber and Integrated Optics.* Vol 21, No 2, 115-124. (2002).
10. A. Biswas. "Dispersion-Managed solitons in optical fibers" *Journal of Optics A.* Vol 4, No 1, 84-97. (2002).
11. A. Biswas. "Integro-differential perturbation of dispersion-managed solitons". *Journal of Electromagnetic Waves and Applications.* Vol 17, No 4, 641-665. (2003).
12. A. Biswas. "Dispersion-managed solitons in optical couplers". *Journal of Nonlinear Optical Physics and Materials.* Vol 12, No 1, 45-74. (2003).
13. A. Biswas. "Perturbations of dispersion-managed optical solitons". *Progress in Electromagnetics Research.* Vol 48, 85-123. (2004).
14. A. Biswas. "Stochastic perturbation of dispersion-managed optical solitons". *Optical and Quantum Electronics.* Vol 37, No 7, 649-659. (2005).

15. A. Biswas & S. Konar. *Introduction to Non-Kerr Law Optical Solitons.* CRC Press, Boca Raton, FL. USA. (2006).
16. D. Breuer, F. Kuppers, A. Mattheus, E. G. Shapiro, I. Gabitov & S. K. Turitsyn. "Symmetrical dispersion compensation for standard monomode fiber based communication systems with large amplifier spacing". *Optics Letters.* Vol 22, No 13, 982-984. (1997).
17. D. Breuer, K. Jürgensen, F. Küppers, A. Mattheus, I. Gabitov & S. K. Turitsyn. "Optimal schemes for dispersion compensation of standard monomode fiber based links". *Optics Communications.* Vol 40, No 1-3, 15-18. (1997).
18. A. Ehrhardt, M. Eiselt, G. Goßkopf, L. Küller, W. Pieper, R. Schnabel, G. H. Weber. "Semiconductor laser amplifier as optical switching gate". *Journal of Lightwave Technology.* Vol 11, No 8, 1287-1295. (1993).
19. A. D. Fishman, G. D. Duff & A. J. Nagel. "Measurement and simulation of multipath interference for 1.7 Gb/s lightwave transmission systems using single- and multifrequency laser". *Journal of Lightwave Technology.* Vol 8, No 6, 894-905. (1990).
20. C. M. Khalique & A. Biswas. "A Lie symmetry approach to nonlinear Schrödinger's equation with non-Kerr law nonlinearity". *Communications in Nonlinear Science and Numerical Simulation.* Vol 14, No 12, 4033-4040. (2009).
21. R. Kohl, A. Biswas, D. Milovic & E. Zerrad. "Perturbation of Gaussian optical solitons in dispersion-managed fibers". *Applied Mathematics and Computation.* Vol 199, No 1, 250-258. (2009).
22. R. Kohl, A. Biswas, D. Milovic & E. Zerrad. "Perturbation of super-sech solitons in dispersion-managed optical fibers". *International Journal of Theoretical Physics.* Vol 47, No 7, 2038-2064. (2008).
23. R. Kohl, A. Biswas, D. Milovic & E. Zerrad. "Adiabatic dynamics of Gaussian and super-Gaussian solitons in dispersion-managed optical fibers". *Progress in Electromagnetics Research.* Vol 84, 27-53. (2008).
24. R. Kohl, D. Milovic, E. Zerrad & A. Biswas. "Perturbation of super-Gaussian optical solitons in dispersion-managed fibers". *Mathematical and Computer Modelling.* Vol 49, No 7-8, 418, 427. (2009).
25. R. Kohl, D. Milovic, E. Zerrad & A. Biswas. "Soliton perturbation theory for dispersion-managed optical fibers". *Journal of Nonlinear Optical Physics and Materials.* Vol 18, No 2. (2009).
26. P. M. Lushnikov. "Dispersion-managed solitons in optical fibers with zero average dispersion". *Optics Letters.* Vol 25, No 16, 1144-1146. (2000).
27. M. F. Mahmood & S. B. Qadri. "Modeling propagation of chirped solitons in an elliptically low birefringent single-mode optical fiber". *Journal of Nonlinear Optical Physics and Materials.* Vol 8, No 4, 469-475. (1999).
28. M. Matsumoto. "Analysis of interaction between stretched pulses propagating in dispersion-managed fibers". *IEEE Photonics Technology Letters.* Vol 10, No 3, 373-375. (1998).
29. H. Michinel. "Pulses nonlinear surface waves and soliton emission at nonlinear graded index waveguides". *Optical and Quantum Electronics.* Vol 30, No 2, 79-97. (1998).
30. A. Panajotovic, D. Milovic & A. Mittic. "Boundary case of pulse propagation analytic solution in the presence of interference and higher order dispersion". *TELSIKS 2005 Conference Proceedings.* 547-550. Nis-Serbia. (2005).
31. A. Panajotovic, D. Milovic, A. Biswas & E. Zerrad. "Influence of even order dispersion on super-sech soliton transmission quality under coherent crosstalk". *Research Letters in Optics.* Vol 2008, 613986, 5 pages. (2008).
32. A. Panajotovic, D. Milovic & A. Biswas. "Influence of even order dispersion on soliton transmission quality with coherent interference". *Progress in Electromagnetics Research B.* Vol 3, 63-72. (2008).
33. V. N. Serkin & A. Hasegawa. "Soliton management in the nonlinear Schrodinger equation model with varying dispersion, nonlinearity and gain". *JETP Letters.* Vol 72, No 2, 89-92. (2000).

34. O. V. Sinkin, V. S. Grigoryan & C. R. Menyuk. "Accurate probabilistic treatment of bit-pattern-dependent nonlinear distortions in BER calculations for WDM RZ systems". *Journal of Lightwave Technology.* Vol 25, No 10, 2959-2967. (2007).
35. M. Stefanovic & D. Milovic. "The impact of out-of-band crosstalk on optical communication link preferences". *Journal of Optical Communications.* Vol 26, No 2, 69-72. (2005).
36. M. Stefanovic, D. Draca, A. Panajotovic & D. Milovic. "Individual and joint influence of second and third order dispersion on transmission quality in the presence of coherent interference". *Optik.* Vol 120, No 13, 636-641.(2009).
37. B. Stojanovic, D. M. Milovic & A. Biswas. "Timing shift of optical pulses due to interchannel crosstalk". *Progress in Electromagnetics Research M.* Vol 1, 21-30. (2008).
38. S. K. Turitsyn, I. Gabitov, E. W. Laedke, V. K. Mezentsev, S. L. Musher, E. G. Shapiro, T. Schäfer, & K. H. Spatschek. "Variational approach to optical pulse propagation in dispersion compensated transmission system". *Optics Communications.* Vol 151, No 1-3, 117-135. (1998).
39. P. K. A. Wai & C. R. Menyuk. "Polarization mode dispersion, de-correlation, and diffusion in optical fibers with randomly varying birefringence". *Journal of Lightwave Technology.* Vol 14, No 2, 148-157. (1996).
40. V. E. Zakharov & S. Wabnitz. *Optical Solitons: Theoretical Challenges and Industrial Perspectives.* Springer, Heidelberg. DE. (1999).

Chapter 4
Birefringent Fibers

4.1 Introduction

An ideal isotropic fiber propagates undisturbed in any state of polarization launched into the fiber. Under ideal conditions of perfect cylindrical geometry and isotropic material, a mode excited with its polarization in one direction would not couple with the mode in the orthogonal direction. Real fibers possess some amount of anisotropy because of an accidental loss of circular symmetry. This loss is due to either a non-circular geometry of the fiber or a non-symmetrical stress field in the fiber cross section. Thus, small deviations from the cylindrical geometry or small fluctuations in material anisotropy result in a mixing of the two polarization states and the mode degeneracy is broken. Thus, the mode propagation constant becomes slightly different for the modes polarized in orthogonal directions. This property is referred to as *modal birefringence* [35].

The intrinsic birefringence is introduced in the manufacturing process,which is a permanent feature of the fiber. It comprises any effect that causes a deviation from the perfect rotational symmetry of the ideal fiber. An elliptical core gives rise to the geometrical birefringence, while a non-symmetrical stress field in the fiber cross-section induces the stress birefringence which is introduced by the photo-elastic effect during the fiber manufacturing process [35].

The birefringence can also be created whenever a fiber undergoes elastic stresses due to external perturbations like the hydrostatic pressure, longitudinal strain, squeezing, twisting, bending and other such external factors. The perturbation induced in the permittivity tensor through the photo-elastic effect lifts the degeneracy of the linearly polarized modes and induces extrinsic birefringence [35].

The birefringence can also significantly affect the soliton propagation in optical fibers. In a highly birefringent polarization maintaining fiber, the difference of phase velocities between the two orthogonally polarized modes is significantly high to avoid coupling between these two modes. In highly birefringent fibers, the beat length (inverse of intrinsic birefringence) is much lower than the nonlinear length and solitons remain stable whether launched close to the slow or fast axis. On the contrary, for weakly birefringent fibers solitons remain stable along the slow axis but become unstable along the fast axis. Under certain conditions the two orthogonally polarized solitons can trap one another and move at a common group velocity, despite the polarization dispersion. This phenomenon is known as *soliton trapping* and can be applied in soliton dragging logic gates. Other applications of nonlinear birefringence include high-resolution distributed fiber sensing and passive mode-locking of fiber lasers [35].

The propagation of solitons in birefringent nonlinear fibers has attracted much attention in recent years. The equations that describe the pulse propagation through these fibers was originally derived by Menyuk [11]. They can be solved approximately in certain special cases only. The localized pulse evolution in a birefringent fiber has been studied analytically, numerically and experimentally [11] on the basis of a simplified chirp-free model without oscillating terms under the assumptions that the two polarizations exhibit different group velocities. In this chapter, the equations that describe the pulse propagation in birefringent fibers are of the following dimensionless form:

$$i(u_z + \delta u_t) + \beta u + \frac{D(z)}{2} u_{tt} + g(z)(|u|^2 + \alpha |v|^2)u + \gamma v^2 u^* = 0 \tag{4.1}$$

$$i(v_z - \delta v_t) + \beta v + \frac{D(z)}{2} v_{tt} + g(z)(|v|^2 + \alpha |u|^2)v + \gamma u^2 v^* = 0 \tag{4.2}$$

Equations (4.1) and (4.2) are known as the Dispersion Managed Vector Nonlinear Schrodinger's Equation (DM-VNLSE). Here, u and v are slowly varying envelopes of the two linearly polarized components of the field along the x and y axis. Also, δ is the group velocity mismatch between the two polarization components and is called the birefringence parameter, β corresponds to the difference between the propagation constants, α is the cross-phase modulation coefficient and γ is the coefficient of the coherent energy coupling term. These equations are, in general, not integrable. However, they can be solved analytically only for certain specific cases [5, 11].

In this analysis, the terms with δ will be neglected as $\delta \leq 10^{-3}$ [5]. Also, neglecting β and the coherent energy coupling given by the coefficient of the γ term, one arrives at the special case of Eqs.(4.1) and (4.2) as

$$iu_z + \frac{D(z)}{2} u_{tt} + g(z)(|u|^2 + \alpha |v|^2)u = 0 \tag{4.3}$$

$$iv_z + \frac{D(z)}{2} v_{tt} + g(z)(|v|^2 + \alpha |u|^2)v = 0 \tag{4.4}$$

4.2 Integrals of motion

The two integrals of the motion of Eqs.(4.1) and (4.2) are the energy (E) and the momentum (M) of the pulse that are respectively given by [11]

$$E = \int_{-\infty}^{\infty} (|u|^2 + |v|^2) dt \tag{4.5}$$

$$M = \frac{i}{2} D(z) \int_{-\infty}^{\infty} (u^* u_t - u u_t^* + v^* v_t - v v_t^*) dt \tag{4.6}$$

By Noether's theorem [5], each of these two conserved quantities corresponds to a symmetry of the system. The conservation of energy is a result of the translational invariance of Eqs.(4.1) and (4.2) relative to phase shifts, while the conservation of the momentum is a consequence of the translational invariance in t [5]. The Hamiltonian (H) given by

$$\begin{aligned} H = \int_{-\infty}^{\infty} \Bigg[& \frac{D(z)}{2} (|u_t|^2 + |v_t|^2) - \beta(|u|^2 - |v|^2) \\ & - \frac{g(z)}{2} (|u|^4 + |v|^4) - i\frac{\delta}{2} (u^* u_t - u u_t^* + v^* v_t - v v_t^*) \\ & - \alpha |u|^2 |v|^2 - \frac{1}{2} (1 - \alpha)(u^2 v^{*2} + v^2 u^{*2}) \Bigg] dt \end{aligned} \tag{4.7}$$

as in the case of polarization preserving fibers, is not a constant of motion, unless $D(z)$ and $g(z)$ are constants in which it is a consequence of the translational invariance in z. For Eqs.(4.3) and (4.4), the Hamiltonian is given by

$$\begin{aligned} H = \int_{-\infty}^{\infty} \Bigg[& \frac{D(z)}{2} (|u_t|^2 + |v_t|^2) - \frac{g(z)}{2} (|u|^4 + |v|^4) \\ & - \alpha |u|^2 |v|^2 - \frac{1}{2} (1 - \alpha)(u^2 v^{*2} + v^2 u^{*2}) \Bigg] dt \end{aligned} \tag{4.8}$$

Now, it is assumed that the solutions of Eqs.(4.3) and (4.4) are given by chirped pulses of the form [5, 11]

$$\begin{aligned} u(z,t) = & A_1(z) f \left[B_1(z) \{ t - t_1(z) \} \right] \exp [i C_1(z) \{ t - t_1(z) \}^2 \\ & - i \kappa_1(z) \{ t - t_1(z) \} + i \theta_1(z)] \end{aligned} \tag{4.9}$$

and

$$\begin{aligned} v(z,t) = & A_2(z) f \left[B_2(z) \{ t - t_2(z) \} \right] \exp [i C_2(z) \{ t - t_2(z) \}^2 \\ & - i \kappa_2(z) \{ t - t_2(z) \} + i \theta_2(z)] \end{aligned} \tag{4.10}$$

where f represents the shape of the pulse. Also, here the parameters $A_j(z)$, $B_j(z)$, $C_j(z)$, $\kappa_j(z)$, $t_j(z)$ and $\theta_j(z)$ (for $j = 1, 2$) represent the pulse amplitudes, the inverse width of the pulses, chirps, frequencies, the centers of the pulses and the phases of the pulses, respectively. Once again, this approach is only approximate and does not account for characteristics such as energy loss due to continuum radiation, damping of the amplitude oscillations and changing of the pulse shape. Also, for such pulse forms, the integrals of motion are

$$E = \int_{-\infty}^{\infty} \left(|u|^2 + |v|^2\right) dt = \frac{A_1^2}{B_1} I_{0,2,0}^{(1)} + \frac{A_2^2}{B_2} I_{0,2,0}^{(2)} \tag{4.11}$$

$$\begin{aligned} M &= \frac{i}{2} D(z) \int_{-\infty}^{\infty} (u^* u_t - u u_t^* + v^* v_t - v v_t^*) dt \\ &= -D(z) \left(\kappa_1 \frac{A_1^2}{B_1} I_{0,2,0}^{(1)} + \kappa_2 \frac{A_2^2}{B_2} I_{0,2,0}^{(2)} \right) \end{aligned} \tag{4.12}$$

The Hamiltonian is now given by

$$\begin{aligned} H = &\int_{-\infty}^{\infty} \left[\frac{D(z)}{2} \left(|u_t|^2 + |v_t|^2\right) - \frac{g(z)}{2} \left(|u|^4 + |v|^4\right) \right. \\ &\left. - \alpha |u|^2 |v|^2 - \frac{1}{2}(1-\alpha) \left(u^2 v^{*2} + v^2 u^{*2}\right) \right] dt \\ = &\frac{D(z)}{2} \left(A_1^2 B_1 I_{0,0,2}^{(1)} + 4 \frac{A_1^2 C_1^2}{B_1^3} I_{2,2,0}^{(1)} + \frac{\kappa^2 A_1^2}{B_1} I_{0,2,0}^{(1)} \right) \\ &- \frac{g(z)}{2} \frac{A_1^4}{B_1} I_{0,4,0}^{(1)} \\ &+ \frac{D(z)}{2} \left(A_2^2 B_2 I_{0,0,2}^{(2)} + 4 \frac{A_2^2 C_2^2}{B_2^3} I_{2,2,0}^{(2)} + \frac{\kappa^2 A_2^2}{B_2} I_{0,2,0}^{(2)} \right) \\ &- \frac{g(z)}{2} \frac{A_2^4}{B_2} I_{0,4,0}^{(2)} - \alpha A_1^2 A_2^2 I + (1-\alpha) A_1 A_2 J \end{aligned} \tag{4.13}$$

where the following notations are introduced:

$$I = \int_{-\infty}^{\infty} f^2 \left[B_1(z) \left(t - t_1(z)\right)\right] f^2 \left[B_2(z) \left(t - t_2(z)\right)\right] dt \tag{4.14}$$

$$\begin{aligned} J = &\int_{-\infty}^{\infty} f \left[B_1(z) \left(t - t_1(z)\right)\right] f \left[B_2(z) \left(t - t_2(z)\right)\right] \cos[C_1 (t - t_1)^2 \\ &- C_2 \left(t - t_2\right)^2 + \kappa_2 (t - t_2) - \kappa_1 (t - t_1) + (\theta_1 - \theta_2)] dt \end{aligned} \tag{4.15}$$

and

$$I_{a,b,c}^{(j)} = \int_{-\infty}^{\infty} \tau_j^a f^b(\tau_j) \left(\frac{df}{d\tau_j}\right)^c d\tau_j \tag{4.16}$$

for $j = 1, 2$.

4.2.1 Gaussian pulses

The conserved quantities, for Gaussian pulses in a birefringent fiber, reduce to

$$E = \int_{-\infty}^{\infty} (|u|^2 + |v|^2)dt = \left(\frac{A_1^2}{B_1} + \frac{A_2^2}{B_2}\right)\sqrt{\frac{\pi}{2}} \tag{4.17}$$

$$\begin{aligned} M &= \frac{i}{2}D(z)\int_{-\infty}^{\infty} (u^* u_t - u u_t^* + v^* v_t - v v_t^*)dt \\ &= -D(z)\left(\kappa_1\frac{A_1^2}{B_1} + \kappa_2\frac{A_2^2}{B_2}\right)\sqrt{\frac{\pi}{2}} \end{aligned} \tag{4.18}$$

while the Hamiltonian is given by

$$\begin{aligned} H = &\int_{-\infty}^{\infty} \left[\frac{D(z)}{2}(|u_t|^2 + |v_t|^2) - \frac{g(z)}{2}(|u|^4 + |v|^4)\right. \\ &\left. - \alpha|u|^2|v|^2 - \frac{1}{2}(1-\alpha)(u^2 v^{*2} + v^2 u^{*2})\right] dt \\ = &\frac{D(z)A_1^4}{2B_1^2}\left(B_1^2 + \frac{C_1^2}{B_1^2} + \kappa_1^2\right)\sqrt{\frac{\pi}{2}} - g(z)B_1 K_1^4\frac{\sqrt{\pi}}{4} \\ &+ \frac{D(z)A_2^4}{B_2^2}\left(B_2^2 + \frac{C_2^2}{B_2^2} + \kappa_2\right)\sqrt{\frac{\pi}{2}} - g(z)B_2 K_2^4\frac{\sqrt{\pi}}{4} \\ &- \alpha A_1^2 A_2^2\sqrt{\frac{\pi}{B_1^2 + B_2^2}}e^{-\frac{B_1^2 B_2^2}{B_1^2 + B_2^2}(t_1 - t_2)^2} + (1-\alpha)A_1 A_2 J \end{aligned} \tag{4.19}$$

where the quantities I and J, for the case of Gaussian pulses, respectively reduced to

$$I = \int_{-\infty}^{\infty} e^{-\left[B_1^2(t-t_1)^2 + B_2^2(t-t_2)^2\right]}dt = \sqrt{\frac{\pi}{B_1^2 + B_2^2}}e^{-\frac{B_1^2 B_2^2}{B_1^2 + B_2^2}(t_1 - t_2)^2} \tag{4.20}$$

and

$$\begin{aligned} J = &\int_{-\infty}^{\infty} e^{-\left[B_1^2(t-t_1)^2 + B_2^2(t-t_2)^2\right]}\cos[C_1(t-t_1)^2 - C_2(t-t_2)^2 \\ &+ \kappa_2(t-t_2) - \kappa_1(t-t_1) + (\theta_1 - \theta_2)]dt \end{aligned} \tag{4.21}$$

4.2.2 Super-Gaussian pulses

The conserved quantities for SG pulses are

$$E = \int_{-\infty}^{\infty} (|u|^2 + |v|^2) dt = \frac{D(z)}{m 2^{\frac{1}{2m}}} \left(\frac{A_1^2}{B_1} + \frac{A_2^2}{B_2} \right) \Gamma \left(\frac{1}{2m} \right) \tag{4.22}$$

$$\begin{aligned} M &= \frac{i}{2} D(z) \int_{-\infty}^{\infty} (u^* u_t - u u_t^* + v^* v_t - v v_t^*) dt \\ &= -\frac{D(z)}{m 2^{\frac{1}{2m}}} \left(\kappa_1 \frac{A_1^2}{B_1} + \kappa_2 \frac{A_2^2}{B_2} \right) \Gamma \left(\frac{1}{2m} \right) \end{aligned} \tag{4.23}$$

while the Hamiltonian here is

$$\begin{aligned} H = & \int_{-\infty}^{\infty} \left[\frac{D(z)}{2} (|u_t|^2 + |v_t|^2) - \frac{g(z)}{2} (|u|^4 + |v|^4) \right. \\ & \left. - \alpha |u|^2 |v|^2 - \frac{1}{2} (1-\alpha)(u^2 v^{*2} + v^2 u^{*2}) \right] dt \\ = & \frac{D(z) A_1^4}{B_1^2} \left[B_1^2 \frac{m}{2^{\frac{2m-1}{2m}}} \Gamma \left(\frac{4m-1}{2m} \right) + \frac{C_1^2}{B_1^2} \frac{2^{\frac{2m-3}{2m}}}{m} \Gamma \left(\frac{3}{2m} \right) \right. \\ & \left. + \kappa_1^2 \frac{1}{m 2^{\frac{2m+1}{m}}} \Gamma \left(\frac{1}{2m} \right) \right] \\ & + \frac{D(z) A_2^4}{B_2^2} \left[B_2^2 \frac{m}{2^{\frac{2m-1}{2m}}} \Gamma \left(\frac{4m-1}{2m} \right) + \frac{C_2^2}{B_2^2} \frac{2^{\frac{2m-3}{2m}}}{m} \Gamma \left(\frac{3}{2m} \right) \right. \\ & \left. + \kappa_2^2 \frac{1}{m 2^{\frac{2m+1}{m}}} \Gamma \left(\frac{1}{2m} \right) \right] - g(z)(B_1 K_1^4 + B_2 K_2^4) \frac{1}{m 2^{\frac{2m+1}{m}}} \Gamma \left(\frac{1}{m} \right) \\ & - \alpha A_1^2 A_2^2 I - (1-\alpha) A_1 A_2 J \end{aligned} \tag{4.24}$$

where the quantities I and J, for SG pulses, respectively reduce to

$$I = \int_{-\infty}^{\infty} e^{-\left[B_1^{2m} (t-t_1)^{2m} + B_2^{2m} (t-t_2)^{2m} \right]} dt \tag{4.25}$$

and

$$\begin{aligned} J = & \int_{-\infty}^{\infty} e^{-\left[B_1^{2m} (t-t_1)^{2m} + B_2^{2m} (t-t_2)^{2m} \right]} \cos[C_1 (t-t_1)^2 - C_2 (t-t_2)^2 \\ & + \kappa_2 (t-t_2) - \kappa_1 (t-t_1) + (\theta_1 - \theta_2)] dt \end{aligned} \tag{4.26}$$

4.3 Variational principle

Since Eqs.(4.3) and (4.4) can not be solved by the aid of the Inverse Scattering Transform, they will be studied by the aid of the variational principle. This is based on the observation that these equations support the chirped soliton solution whose shape is close to that of a Gaussian or sometimes a Super-

Gaussian(SG) pulse. For DM-VNLSE, the Lagrangian is given by [5]:

$$L = \frac{1}{2}\int_{-\infty}^{\infty} [i\left(u^* u_z - u u_z^*\right) + i\left(v^* v_z - v v_z^*\right) \\ + i\delta\left(v^* u_t - u v_t^*\right) + i\delta\left(u^* v_t - v u_t^*\right) - D(z)(|u_t|^2 + |v_t|^2) \\ + g(z)(|u|^4 + |v|^4) + 2\alpha g(z)|u|^2|v|^2 \\ + 2\beta\left(u^* v + u v^*\right) + \gamma(u^2 v^{*2} + v^2 u^{*2})]dt \tag{4.27}$$

For the reduced DM-VNLSE, the Lagrangian given by (4.3)-(4.4) is given by

$$L = \int_{-\infty}^{\infty} \left[\frac{i}{2}\left(u^* u_z - u u_z^*\right) + \frac{i}{2}\left(v^* v_z - v v_z^*\right) - \frac{D(z)}{2}(|u_t|^2 + |v_t|^2) \right. \\ \left. + \frac{g(z)}{2}(|u|^4 + |v|^4) + \alpha g(z)|u|^2|v|^2\right] dt \tag{4.28}$$

Now, using the pulses given by Eqs.(4.9) and (4.10) the Lagrangian reduces to

$$L = -D(z)A_1^2\left(\frac{B_1}{2}I_{0,0,2}^{(1)} + 2\frac{C_1^2}{B_1^3}I_{2,2,0}^{(1)} + \frac{\kappa_1^2}{2B_1}I_{0,2,0}^{(1)}\right) \\ + \frac{g(z)A_1^4}{2B_1}I_{0,4,0}^{(1)} - \frac{A_1^2}{B_1^3}I_{2,2,0}^{(1)}\frac{dC_1}{dz} + \frac{A_1^2}{B_1}I_{0,2,0}^{(1)}\left(t_1\frac{d\kappa_1}{dz} - \frac{d\theta_1}{dz}\right) \\ - D(z)A_2^2\left(\frac{B_2}{2}I_{0,0,2}^{(2)} + 2\frac{C_2^2}{B_2^3}I_{2,2,0}^{(2)} + \frac{\kappa_2^2}{2B_2}I_{0,2,0}^{(2)}\right) \\ + \frac{g(z)A_2^4}{2B_2}I_{0,4,0}^{(2)} - \frac{A_2^2}{B_2^3}I_{2,2,0}^{(2)}\frac{dC_2}{dz} + \frac{A_2^2}{B_2}I_{0,2,0}^{(2)}\left(t_2\frac{d\kappa_2}{dz} - \frac{d\theta_2}{dz}\right) \\ + \alpha g(z)A_1^2A_2^2 I \tag{4.29}$$

The EL equation given by Eq.(2.82) will be utilized to obtain the dynamics of pulse parameters for birefringent fibers. In EL equation, p now represents one of the twelve soliton parameters. Substituting A_j, B_j, C_j, κ_j, t_j and θ_j $(j = 1, 2)$ for p in (2.82), the following set of equations are obtained:

$$\frac{dA_1}{dz} = -D(z)A_1C_1 \tag{4.30}$$

$$\frac{dB_1}{dz} = -2D(z)B_1C_1 \tag{4.31}$$

$$\frac{dC_1}{dz} = D(z)\left(\frac{B_1^4}{2}\frac{I_{0,0,2}^{(1)}}{I_{2,2,0}^{(1)}} - 2C_1^2\right) - \frac{g(z)A_1^2B_1^2}{4}\frac{I_{0,4,0}^{(1)}}{I_{2,2,0}^{(1)}} \\ - \frac{\alpha g(z)}{2}A_2^2B_1^3\frac{I}{I_{2,2,0}^{(1)}} \tag{4.32}$$

$$\frac{d\kappa_1}{dz} = 0 \tag{4.33}$$

$$\frac{dt_1}{dz} = -D(z)\kappa_1 \tag{4.34}$$

$$\frac{d\theta_1}{dz} = D(z)\left(\frac{\kappa_1^2}{2} - B_1^2 \frac{I_{0,0,2}^{(1)}}{I_{0,2,0}^{(1)}}\right) + \frac{5g(z)A_1^2}{4}\frac{I_{0,4,0}^{(1)}}{I_{0,2,0}^{(1)}} + \frac{3}{2}\alpha g(z) A_2^2 B_1 \frac{I}{I_{0,2,0}^{(1)}} \tag{4.35}$$

$$\frac{dA_2}{dz} = -D(z)A_2C_2 \tag{4.36}$$

$$\frac{dB_2}{dz} = -2D(z)B_2C_2 \tag{4.37}$$

$$\frac{dC_2}{dz} = D(z)\left(\frac{B_2^4}{2}\frac{I_{0,0,2}^{(2)}}{I_{2,2,0}^{(2)}} - 2C_2^2\right) - \frac{g(z)A_2^2B_2^2}{4}\frac{I_{0,4,0}^{(2)}}{I_{2,2,0}^{(2)}} - \frac{\alpha g(z)}{2} A_1^2 B_2^3 \frac{I}{I_{2,2,0}^{(2)}} \tag{4.38}$$

$$\frac{d\kappa_2}{dz} = 0 \tag{4.39}$$

$$\frac{dt_2}{dz} = -D(z)\kappa_2 \tag{4.40}$$

$$\frac{d\theta_2}{dz} = D(z)\left(\frac{\kappa_2^2}{2} - B_2^2 \frac{I_{0,0,2}^{(2)}}{I_{0,2,0}^{(2)}}\right) + \frac{5g(z)A_2^2}{4}\frac{I_{0,4,0}^{(2)}}{I_{0,2,0}^{(2)}} + \frac{3}{2}\alpha g(z) A_1^2 B_2 \frac{I}{I_{0,2,0}^{(2)}} \tag{4.41}$$

The explicit form of the parameter dynamics for the Gaussian and SG pulses will now be obtained in the following two subsections.

4.3.1 Gaussian pulses

The evolution equations (4.30)–(4.41) respectively reduce to

$$\frac{dA_1}{dz} = -D(z)A_1C_1 \tag{4.42}$$

$$\frac{dB_1}{dz} = -2D(z)B_1C_1 \tag{4.43}$$

$$\frac{dC_1}{dz} = 2D(z)(B_1^4 - C_1^2) - \frac{\sqrt{2}}{2}g(z)A_1^4B_1 - \frac{2\alpha g(z)A_2^4B_1^3}{B_2}\sqrt{\frac{2}{B_1^2+B_2^2}}e^{-\frac{B_1^2B_2^2}{B_1^2+B_2^2}(t_1-t_2)^2} \tag{4.44}$$

$$\frac{d\kappa_1}{dz} = 0 \tag{4.45}$$

$$\frac{dt_1}{dz} = -D(z)\kappa_1 \tag{4.46}$$

$$\frac{d\theta_1}{dz} = \frac{5}{4\sqrt{2}}\frac{g(z)A_1^4}{B_1} + \frac{D(z)}{2}(\kappa_1^2 - 2B_1^2) + \frac{3\alpha g(z)A_2^4B_1^3}{2B_2}\sqrt{\frac{2}{B_1^2+B_2^2}}e^{-\frac{B_1^2B_2^2}{B_1^2+B_2^2}(t_1-t_2)^2} \tag{4.47}$$

$$\frac{dA_2}{dz} = -D(z)A_2C_2 \tag{4.48}$$

$$\frac{dB_2}{dz} = -2D(z)B_2C_2 \tag{4.49}$$

$$\frac{dC_2}{dz} = 2D(z)(B_2^4 - C_2^2) - \frac{\sqrt{2}}{2}g(z)A_2^4B_2 - \frac{2\alpha g(z)A_1^4B_2^3}{B_1}\sqrt{\frac{2}{B_1^2+B_2^2}}e^{-\frac{B_1^2B_2^2}{B_1^2+B_2^2}(t_1-t_2)^2} \tag{4.50}$$

$$\frac{d\kappa_2}{dz} = 0 \tag{4.51}$$

$$\frac{dt_2}{dz} = -D(z)\kappa_2 \tag{4.52}$$

$$\frac{d\theta_2}{dz} = \frac{5}{4\sqrt{2}}\frac{g(z)A_2^4}{B_2} + \frac{D(z)}{2}(\kappa_2^2 - 2B_2^2) + \frac{3\alpha g(z)A_1^4B_2}{2B_1}\sqrt{\frac{2}{B_1^2+B_2^2}}e^{-\frac{B_1^2B_2^2}{B_1^2+B_2^2}(t_1-t_2)^2} \tag{4.53}$$

These equations are useful in studying the various physical aspects of the solitons in birefringent optical fibers, namely, the timing, amplitude or the frequency jitter, the evolution of the coherent energy and much more.

4.3.2 Super-Gaussian pulses

In this case, the evolution equations (4.30)–(4.41) respectively reduce to

$$\frac{dA_1}{dz} = -D(z)A_1C_1 \tag{4.54}$$

$$\frac{dB_1}{dz} = -2D(z)B_1C_1 \tag{4.55}$$

$$\frac{dC_1}{dz} = D(z)\left\{\frac{m^2}{2^{\frac{m-2}{m}}}\frac{\Gamma\left(\frac{4m-1}{2m}\right)}{\Gamma\left(\frac{3}{2m}\right)}B^4 - 2C^2\right\}$$
$$- g(z)A_1^4B_1\frac{1}{2^{\frac{4m+1}{2m}}}\frac{\Gamma\left(\frac{1}{2m}\right)}{\Gamma\left(\frac{3}{2m}\right)} - \frac{m}{2^{\frac{2m-3}{m}}}\frac{\alpha g(z)A_2^4}{B_2}\frac{I}{\Gamma\left(\frac{3}{2m}\right)} \tag{4.56}$$

$$\frac{d\kappa_1}{dz} = 0 \tag{4.57}$$

$$\frac{dt_1}{dz} = -D(z)\kappa_1 \tag{4.58}$$

$$\frac{d\theta_1}{dz} = D(z)\left\{\frac{\kappa^2}{2} - m^2 2^{\frac{1}{m}}\frac{\Gamma\left(\frac{4m-1}{2m}\right)}{\Gamma\left(\frac{1}{2m}\right)}B_1^2\right\}$$
$$+ \frac{5}{2^{\frac{4m+1}{2m}}}\frac{g(z)A_1^2}{B_1} + \frac{3m}{2^{\frac{2m-1}{2m}}}\frac{\alpha g(z)A_2^4}{B_2}\frac{I}{\Gamma\left(\frac{1}{2m}\right)} \tag{4.59}$$

$$\frac{dA_2}{dz} = -D(z)A_2C_2 \tag{4.60}$$

$$\frac{dB_2}{dz} = -2D(z)B_2C_2 \tag{4.61}$$

$$\frac{dC_2}{dz} = D(z)\left\{\frac{m^2}{2^{\frac{m-2}{m}}}\frac{\Gamma\left(\frac{4m-1}{2m}\right)}{\Gamma\left(\frac{3}{2m}\right)}B^4 - 2C^2\right\}$$
$$- g(z)A_2^4B_2\frac{1}{2^{\frac{4m+1}{2m}}}\frac{\Gamma\left(\frac{1}{2m}\right)}{\Gamma\left(\frac{3}{2m}\right)} - \frac{m}{2^{\frac{2m-3}{m}}}\frac{\alpha g(z)A_1^4B_2^3}{B_1}\frac{I}{\Gamma\left(\frac{3}{2m}\right)} \tag{4.62}$$

$$\frac{d\kappa_2}{dz} = 0 \tag{4.63}$$

$$\frac{dt_2}{dz} = -D(z)\kappa_2 \tag{4.64}$$

$$\frac{d\theta_2}{dz} = D(z)\left\{\frac{\kappa^2}{2} - m^2 2^{\frac{1}{m}}\frac{\Gamma\left(\frac{4m-1}{2m}\right)}{\Gamma\left(\frac{1}{2m}\right)}B_2^2\right\}$$
$$+ \frac{5}{2^{\frac{4m+1}{2m}}}g(z)A_2^2 + \frac{3m}{2^{\frac{2m-1}{2m}}}\frac{\alpha g(z)A_1^4B_2}{B_1}\frac{I}{\Gamma\left(\frac{1}{2m}\right)} \tag{4.65}$$

4.4 Perturbation terms

In this section, DM-VNLSE in presence of the perturbation terms will be studied. The perturbed DM-VNLSE that are going to be analyzed are given by

$$iu_z + \frac{D(z)}{2}u_{tt} + g(z)(|u|^2 + \alpha|v|^2)u = i\epsilon R_1[u, u^*; v, v^*] \tag{4.66}$$

$$iv_z + \frac{D(z)}{2}v_{tt} + g(z)(|v|^2 + \alpha|u|^2)v = i\epsilon R_2[v, v^*; u, u^*] \tag{4.67}$$

Here, R_1 and R_2 represent the perturbation terms and the perturbation parameter ϵ, as before, is the relative width of the spectrum. In presence of the perturbation terms, the EL equations modify to [27]

$$\frac{\partial L}{\partial p} - \frac{d}{dz}\left(\frac{\partial L}{\partial p_z}\right) = i\epsilon \int_{-\infty}^{\infty} \left(R_1 \frac{\partial u^*}{\partial p} - R_1^* \frac{\partial u}{\partial p}\right) dt \tag{4.68}$$

and

$$\frac{\partial L}{\partial p} - \frac{d}{dz}\left(\frac{\partial L}{\partial p_z}\right) = i\epsilon \int_{-\infty}^{\infty} \left(R_2 \frac{\partial v^*}{\partial p} - R_2^* \frac{\partial v}{\partial p}\right) dt \tag{4.69}$$

where p represents twelve soliton parameters. Once again, substituting A_j, B_j, C_j, κ_j, t_j and θ_j, where $j = 1, 2$, for p in (4.68) and (4.69), the following adiabatic evolution equations are obtained:

$$\frac{dA_1}{dz} = -D(z)A_1C_1 - \frac{\epsilon}{2I^{(1)}_{0,2,0}I^{(1)}_{2,2,0}} \int_{-\infty}^{\infty} \Re[R_1 e^{-i\phi_2}] \cdot (I^{(1)}_{0,2,0}\tau_1^2 - 3I^{(1)}_{2,2,0}) f(\tau_1) d\tau_1 \tag{4.70}$$

$$\frac{dB_1}{dz} = -2D(z)B_1C_1 - \frac{\epsilon B_1}{A_1 I^{(1)}_{0,2,0}I^{(1)}_{2,2,0}} \int_{-\infty}^{\infty} \Re[R_1 e^{-i\phi_1}] \cdot (I^{(1)}_{0,2,0}\tau_1^2 - I^{(1)}_{2,2,0}) f(\tau_1) d\tau_1 \tag{4.71}$$

$$\frac{dC_1}{dz} = D(z)\left(\frac{B_1^4}{2}\frac{I^{(1)}_{0,0,2}}{I^{(1)}_{2,2,0}} - 2C_1^2\right) \frac{g(z)A_1^2B_1^2}{4}\frac{I_{0,4,0}}{I_{2,2,0}} - \frac{\alpha g(z)}{2}A_2^2B_1^3\frac{I}{I^{(1)}_{2,2,0}} - \frac{\epsilon B_1^2}{2A_1 I^{(1)}_{2,2,0}} \int_{-\infty}^{\infty} \Im[R_1 e^{-i\phi_2}] \left(f(\tau_1) + 2\tau_1 \frac{df}{d\tau_1}\right) d\tau_1 \tag{4.72}$$

$$\frac{d\kappa_1}{dz} = \frac{2\epsilon}{A_1 B_1 I^{(1)}_{0,2,0}} \int_{-\infty}^{\infty} \left\{ B_1^2 \Im[R_1 e^{-i\phi_1}] \frac{df}{d\tau_1} - 2C_1 \Re[R_1 e^{-i\phi_1}] \tau_1 f(\tau_1) \right\} d\tau_1 \tag{4.73}$$

$$\frac{dt_1}{dz} = -D(z)\kappa_1 + \frac{2\epsilon}{A_1 B_1 I^{(1)}_{0,2,0}} \int_{-\infty}^{\infty} \Re[R_1 e^{-i\phi_1}]\tau_2 f(\tau_1) d\tau_1 \tag{4.74}$$

$$\frac{d\theta_1}{dz} = D(z)\left(\frac{\kappa_1^2}{2} - \frac{I^{(1)}_{0,0,2}}{I^{(1)}_{0,2,0}} B_1^2\right) + \frac{5g(z)A_1^2}{4}\frac{I^{(1)}_{0,4,0}}{I^{(1)}_{0,2,0}} + \frac{3}{2}\alpha g(z) A_2^2 B_1 \frac{I}{I^{(1)}_{0,2,0}}$$
$$+ \frac{\epsilon}{2A_1 B_1 I^{(1)}_{0,2,0}} \int_{-\infty}^{\infty} \left\{ B_1 \Im[R_1 e^{-i\phi_2}]\left(3f(\tau_1) + 2\tau_2 \frac{df}{d\tau_2}\right)\right.$$
$$\left. + 4\kappa_1 \Re[R_1 e^{-i\phi_1}]\tau_1 f(\tau_1)\right\} d\tau_1 \tag{4.75}$$

$$\frac{dA_2}{dz} = -D(z)A_2 C_2 - \frac{\epsilon}{2I^{(2)}_{0,2,0} I^{(2)}_{2,2,0}} \int_{-\infty}^{\infty} \Re[R_2 e^{-i\phi_2}]$$
$$\cdot (I^{(2)}_{0,2,0}\tau_2^2 - 3I^{(2)}_{2,2,0}) f(\tau_2) d\tau_2 \tag{4.76}$$

$$\frac{dB_2}{dz} = -2D(z)B_2 C_2 - \frac{\epsilon B_2}{A_1 I^{(2)}_{0,2,0} I^{(2)}_{2,2,0}} \int_{-\infty}^{\infty} \Re[R_2 e^{-i\phi_2}]$$
$$\cdot (I^{(2)}_{0,2,0}\tau_2^2 - I^{(2)}_{2,2,0}) f(\tau_2) d\tau_2 \tag{4.77}$$

$$\frac{dC_2}{dz} = D(z)\left(\frac{B_2^4}{2}\frac{I^{(2)}_{0,0,2}}{I^{(2)}_{2,2,0}} - 2C_2^2\right) - \frac{g(z)A_2^2 B_2^2}{4}\frac{I^{(2)}_{0,4,0}}{I^{(2)}_{2,2,0}}$$
$$- \frac{\alpha g(z)}{2} A_1^2 B_2^3 \frac{I}{I^{(2)}_{2,2,0}} - \frac{\epsilon B_2^2}{2A_2 I^{(2)}_{2,2,0}} \int_{-\infty}^{\infty} \Im[R_2 e^{-i\phi_2}]$$
$$\cdot \left(f(\tau_2) + 2\tau_2 \frac{df}{d\tau_2}\right) d\tau_2 \tag{4.78}$$

$$\frac{d\kappa_2}{dz} = \frac{2\epsilon}{A_2 B_2 I^{(2)}_{0,2,0}} \int_{-\infty}^{\infty} \left\{ B_2^2 \Im[R_2 e^{-i\phi_1}] \frac{df}{d\tau_2}\right.$$
$$\left. - 2C_2 \Re[R_2 e^{-i\phi_2}]\tau_2 f(\tau_2)\right\} d\tau_2 \tag{4.79}$$

$$\frac{dt_2}{dz} = -D(z)\kappa_2 + \frac{2\epsilon}{A_2 B_2 I^{(2)}_{0,2,0}} \int_{-\infty}^{\infty} \Re[R_2 e^{-i\phi_2}]\tau_2 f(\tau_2) d\tau_2 \tag{4.80}$$

$$\frac{d\theta_2}{dz} = D(z)\left(\frac{\kappa_2^2}{2} - \frac{I^{(2)}_{0,0,2}}{I^{(2)}_{0,2,0}} B_2^2\right) + \frac{5g(z)A_2^2}{4}\frac{I^{(2)}_{0,4,0}}{I^{(2)}_{0,2,0}} + \frac{3}{2}\alpha g A_1^2 B_2 \frac{I}{I^{(2)}_{0,2,0}}$$
$$+ \frac{\epsilon}{2A_2 B_2 I^{(2)}_{0,2,0}} \int_{-\infty}^{\infty} \left\{ B_2 \Im[R_2 e^{-i\phi_2}]\left(3f(\tau_2) + 2\tau_2 \frac{df}{d\tau_2}\right)\right.$$

$$+ 4\kappa_2 \Re[R_2 e^{-i\phi_2}] \tau_2 f(\tau_2) \Big\} d\tau_2 \tag{4.81}$$

where the notations

$$\tau_j = B_j(z)\,(t - t_j(z))$$

and

$$\phi_j = C_j(z)\,\{t - t_j(z)\}^2 - \kappa_j(z)\,\{t - t_j(z)\} + \theta_j(z)$$

for $j = 1, 2$ were used. Once again, $\Re$ and $\Im$ represent the real and imaginary parts, respectively. This dynamics will now be simplified for the Gaussian and SG solitons in the following subsections.

4.4.1 Gaussian pulses

Here, again, using $f(\tau_j) = e^{-\tau_j^2}$ where $j = 1, 2$ and using the integrals $I_{a,b,c}^{(j)}$ for $j = 1, 2$ in Eqs.(4.70)–(4.81), the adiabatic parameter dynamics of perturbed Gaussian pulses are obtained as follows:

$$\frac{dA_1}{dz} = -D(z)A_1C_1 - \frac{\epsilon}{\sqrt{2\pi}} \int_{-\infty}^{\infty} \Re[R_1 e^{-i\phi_1}](4\tau_1^2 - 3)e^{-\tau_1^2} d\tau_1 \tag{4.82}$$

$$\frac{dB_1}{dz} = -2D(z)B_1C_1 - \epsilon\sqrt{\frac{2}{\pi}\frac{B_1}{A_1}} \int_{-\infty}^{\infty} \Re[R_1 e^{-i\phi_1}](4\tau_1^2 - 1)e^{-\tau_1^2} d\tau_1 \tag{4.83}$$

$$\begin{aligned} \frac{dC_1}{dz} &= 2D(z)(B_1^4 - C_1^2) - \frac{1}{\sqrt{2}} g(z) A_1^2 B_1^2 \\ &- 2\alpha g(z) A_2^2 B_1^3 \sqrt{\frac{2}{B_1^2 + B_2^2}}\, e^{-\frac{B_1^2 B_2^2}{B_1^2 + B_2^2}(t_1 - t_2)^2} \\ &- 2\epsilon\sqrt{\frac{2}{\pi}\frac{B_1^2}{A_1}} \int_{-\infty}^{\infty} \Im[R_1 e^{-i\phi_1}](1 - 4\tau_1^2)e^{-\tau_1^2} d\tau_1 \end{aligned} \tag{4.84}$$

$$\begin{aligned} \frac{d\kappa_1}{dz} &= -\frac{2\epsilon}{A_1 B_1}\sqrt{\frac{2}{\pi}} \int_{-\infty}^{\infty} \{B_1^2 \Im[R_1 e^{-i\phi_1}] 2\tau_1 \\ &+ 2C_1 \Re[R_1 e^{-i\phi_1}]\tau_1\} e^{-\tau_1^2} d\tau_1 \end{aligned} \tag{4.85}$$

$$\frac{dt_1}{dz} = -D(z)\kappa_1 + \frac{2\epsilon}{A_1 B_1}\sqrt{\frac{2}{\pi}} \int_{-\infty}^{\infty} \Re[R_1 e^{-i\phi_1}]\tau_1 e^{-\tau_1^2} d\tau_1 \tag{4.86}$$

$$\frac{d\theta_1}{dz} = D(z)\left(\frac{\kappa_1^2}{2} - B_1^2\right) + \frac{5}{4\sqrt{2}} g(z) A_1^2$$

$$+ \frac{3}{2}\alpha g(z)A_2^2B_1\sqrt{\frac{2}{B_1^2+B_2^2}}e^{-\frac{B_1^2B_2^2}{B_1^2+B_2^2}(t_1-t_2)^2}$$

$$+ \frac{\epsilon}{\sqrt{2\pi}}\frac{1}{A_1B_1}\int_{-\infty}^{\infty}\{B_1\Im[R_1e^{-i\phi_1}](3-4\tau_1^2)$$

$$+ 4\kappa_1\Re[R_1e^{-i\phi_1}]\tau_1\}e^{-\tau_1^2}d\tau_1 \tag{4.87}$$

$$\frac{dA_2}{dz} = -D(z)A_2C_2 - \frac{\epsilon}{\sqrt{2\pi}}\int_{-\infty}^{\infty}\Re[R_2e^{-i\phi_2}](4\tau_2^2-3)e^{-\tau_2^2}d\tau_2 \tag{4.88}$$

$$\frac{dB_2}{dz} = -2D(z)B_2C_2 - \epsilon\sqrt{\frac{2}{\pi}}\frac{B_2}{A_2}\int_{-\infty}^{\infty}\Re[R_2e^{-i\phi_2}](4\tau_2^2-1)e^{-\tau_2^2}d\tau_2 \tag{4.89}$$

$$\frac{dC_2}{dz} = 2D(z)\left(B_2^4-C_2^2\right) - \frac{1}{\sqrt{2}}gA_2^2B_2^2$$

$$- 2\alpha g(z)A_1^2B_2^3\sqrt{\frac{2}{B_1^2+B_2^2}}e^{-\frac{B_1^2B_2^2}{B_1^2+B_2^2}(t_1-t_2)^2}$$

$$- 2\epsilon\sqrt{\frac{2}{\pi}}\frac{B_2^2}{A_2}\int_{-\infty}^{\infty}\Im[R_2e^{-i\phi_2}](1-4\tau_2^2)e^{-\tau_2^2}d\tau_2 \tag{4.90}$$

$$\frac{d\kappa_2}{dz} = -\frac{2\epsilon}{A_2B_2}\sqrt{\frac{2}{\pi}}\int_{-\infty}^{\infty}\{B_2^2\Im[R_2e^{-i\phi_2}]2\tau_2$$

$$+ 2C_2\Re[R_2e^{-i\phi_2}]\tau_2\}e^{-\tau_2^2}d\tau_2 \tag{4.91}$$

$$\frac{dt_2}{dz} = -D(z)\kappa_2 + \frac{2\epsilon}{A_2B_2}\sqrt{\frac{2}{\pi}}\int_{-\infty}^{\infty}\Re[R_2e^{-i\phi_2}]\tau_2e^{-\tau_2^2}d\tau_2 \tag{4.92}$$

$$\frac{d\theta_2}{dz} = D(z)\left(\frac{\kappa_2^2}{2}-B_2^2\right) + \frac{5}{4\sqrt{2}}g(z)A_2^2$$

$$+ \frac{3}{2}\alpha g(z)A_1^2B_2\sqrt{\frac{2}{B_1^2+B_2^2}}e^{-\frac{B_1^2B_2^2}{B_1^2+B_2^2}(t_1-t_2)^2}$$

$$+ \frac{\epsilon}{\sqrt{2\pi}}\frac{1}{A_2B_2}\int_{-\infty}^{\infty}\{B_2\Im[R_2e^{-i\phi_2}](3-4\tau_2^2)$$

$$+ 4\kappa_2\Re[R_2e^{-i\phi_2}]\tau_2\}e^{-\tau_2^2}d\tau_2 \tag{4.93}$$

4.4.2 Super-Gaussian pulses

For the SG pulses, the adiabatic parameter dynamics reduces to

$$\frac{dA_1}{dz} = -D(z)A_1C_1 - \epsilon\frac{m^2 2^{\frac{m+2}{m}}}{\Gamma\left(\frac{1}{2m}\right)\Gamma\left(\frac{3}{2m}\right)}\int_{-\infty}^{\infty}\Re[R_1e^{-i\phi_1}]$$

$$\cdot\left\{\frac{\tau_1^2}{m2^{\frac{1}{m}}}\Gamma\left(\frac{1}{2m}\right) - \frac{3}{m2^{\frac{3}{2m}}}\Gamma\left(\frac{3}{2m}\right)\right\}e^{-\tau_1^{2m}}d\tau_1 \tag{4.94}$$

$$\frac{dB_1}{dz} = -2D(z)B_1C_1 - \epsilon\frac{B_1}{A_1}\frac{m^2 2^{\frac{2}{m}}}{\Gamma\left(\frac{1}{2m}\right)\Gamma\left(\frac{3}{2m}\right)}\int_{-\infty}^{\infty}\Re[R_1e^{-i\phi_1}]$$

$$\cdot\left\{\frac{\tau_1^2}{m2^{\frac{1}{2m}}}\Gamma\left(\frac{1}{2m}\right) - \frac{1}{m2^{\frac{3}{2m}}}\Gamma\left(\frac{3}{2m}\right)\right\}e^{-\tau_1^{2m}}d\tau_1 \tag{4.95}$$

$$\frac{dC_1}{dz} = D(z)\left\{B_1^4\frac{m^2}{2^{\frac{m-2}{m}}}\frac{\Gamma\left(\frac{4m-1}{2m}\right)}{\Gamma\left(\frac{3}{2m}\right)} - 2C_1^2\right\}$$

$$- g(z)A_1^2B_1^2\frac{1}{2^{\frac{4m+1}{2m}}}\frac{\Gamma\left(\frac{1}{2m}\right)}{\Gamma\left(\frac{3}{2m}\right)} - \frac{m}{2^{\frac{2m-3}{2m}}}A_2^3B_1^2\frac{I}{\Gamma\left(\frac{3}{2m}\right)}$$

$$- \epsilon\frac{B_1^2}{A_1}\frac{m2^{\frac{2m+3}{2m}}}{\Gamma\left(\frac{3}{2m}\right)}\int_{-\infty}^{\infty}\Im[R_1e^{-i\phi_1}](1-4m\tau_1^{2m})e^{-\tau_1^{2m}}d\tau_1 \tag{4.96}$$

$$\frac{d\kappa_1}{dz} = -\epsilon\frac{1}{A_1B_1}\frac{m2^{\frac{2m-1}{2m}}}{\Gamma\left(\frac{1}{2m}\right)}\int_{-\infty}^{\infty}\{2m\tau_1^{2m-1}B_1^2\Im[R_1e^{-i\phi_1}]$$

$$+ 2\tau C_1\Re[R_1e^{-i\phi_1}]\}e^{-\tau_1^{2m}}d\tau_1 \tag{4.97}$$

$$\frac{dt_1}{dz} = -D(z)\kappa_1 + \epsilon\frac{1}{A_1B_1}\frac{m2^{\frac{2m+1}{2m}}}{\Gamma\left(\frac{1}{2m}\right)}\int_{-\infty}^{\infty}\Re[R_1e^{-i\phi_1}]\tau_1e^{-\tau_1^{2m}}d\tau_1 \tag{4.98}$$

$$\frac{d\theta_1}{dz} = D(z)\left\{\frac{\kappa_1^2}{2} - m^2 2^{\frac{1}{m}}\frac{\Gamma\left(\frac{4m-1}{2m}\right)}{\Gamma\left(\frac{1}{2m}\right)}B_1^2\right\} + 5g(z)A_1^2\frac{1}{2^{\frac{4m+1}{2m}}}$$

$$- \frac{3m}{2^{\frac{2m-1}{2m}}}\alpha g(z)A_2^2B_1\frac{I}{\Gamma\left(\frac{1}{2m}\right)}$$

$$+ \epsilon\frac{1}{A_1B_1}\frac{m2^{\frac{2m+1}{2m}}}{\Gamma\left(\frac{1}{2m}\right)}\int_{-\infty}^{\infty}\{B_1\Im[R_1e^{-i\phi_1}](3-4m\tau_1^{2m})$$

$$+ 4\kappa_1\Re[R_1e^{-i\phi_1}]\tau_1\}e^{-\tau_1^{2m}}d\tau_1 \tag{4.99}$$

$$\frac{dA_2}{dz} = -D(z)A_2C_2 - \epsilon\frac{m^2 2^{\frac{m+2}{m}}}{\Gamma\left(\frac{1}{2m}\right)\Gamma\left(\frac{3}{2m}\right)}\int_{-\infty}^{\infty}\Re[R_2e^{-i\phi_2}]$$

$$\cdot\left\{\frac{\tau_2^2}{m2^{\frac{1}{m}}}\Gamma\left(\frac{1}{2m}\right) - \frac{3}{m2^{\frac{3}{2m}}}\Gamma\left(\frac{3}{2m}\right)\right\}e^{-\tau_2^{2m}}d\tau_2 \tag{4.100}$$

$$\frac{dB_2}{dz} = -2D(z)B_2C_2 - \epsilon\frac{B_2}{A_2}\frac{m^2 2^{\frac{2}{m}}}{\Gamma\left(\frac{1}{2m}\right)\Gamma\left(\frac{3}{2m}\right)}\int_{-\infty}^{\infty}\Re[R_2e^{-i\phi_2}]$$

$$\cdot\left\{\frac{\tau_2^2}{m2^{\frac{1}{2m}}}\Gamma\left(\frac{1}{2m}\right)-\frac{1}{m2^{\frac{3}{2m}}}\Gamma\left(\frac{3}{2m}\right)\right\}e^{-\tau_2^{2m}}d\tau_2 \tag{4.101}$$

$$\frac{dC_2}{dz}=D(z)\left\{B_2^4\frac{m^2}{2^{\frac{m-2}{m}}}\frac{\Gamma\left(\frac{4m-1}{2m}\right)}{\Gamma\left(\frac{3}{2m}\right)}-2C_2^2\right\}$$

$$-g(z)A_2^2B_2^2\frac{1}{2^{\frac{4m+1}{2m}}}\frac{\Gamma\left(\frac{1}{2m}\right)}{\Gamma\left(\frac{3}{2m}\right)}-\frac{m}{2^{\frac{2m-3}{2m}}}A_1^3B_2^2\frac{I}{\Gamma\left(\frac{3}{2m}\right)}$$

$$-\epsilon\frac{B_2^2}{A_2}\frac{m2^{\frac{2m+3}{2m}}}{\Gamma\left(\frac{3}{2m}\right)}\int_{-\infty}^{\infty}\Im[R_2e^{-i\phi_2}](1-4m\tau_2^{2m})e^{-\tau_2^{2m}}d\tau_2 \tag{4.102}$$

$$\frac{d\kappa_2}{dz}=-\epsilon\frac{1}{A_2B_2}\frac{m2^{\frac{2m-1}{2m}}}{\Gamma\left(\frac{1}{2m}\right)}\int_{-\infty}^{\infty}\{2m\tau_2^{2m-1}B_2^2\Im[R_2e^{-i\phi_2}]$$

$$+2\tau_2C_2\Re[R_2e^{-i\phi_2}]\}e^{-\tau_2^{2m}}d\tau_2 \tag{4.103}$$

$$\frac{dt_2}{dz}=-D(z)\kappa_2+\epsilon\frac{1}{A_2B_2}\frac{m2^{\frac{2m+1}{2m}}}{\Gamma\left(\frac{1}{2m}\right)}\int_{-\infty}^{\infty}\Re[R_2e^{-i\phi_2}]\tau_2e^{-\tau_2^{2m}}d\tau \tag{4.104}$$

$$\frac{d\theta_2}{dz}=D(z)\left\{\frac{\kappa_2^2}{2}-m^22^{\frac{1}{m}}\frac{\Gamma\left(\frac{4m-1}{2m}\right)}{\Gamma\left(\frac{1}{2m}\right)}B_2^2\right\}+5g(z)A_2^2\frac{1}{2^{\frac{4m+1}{2m}}}$$

$$-\frac{3m}{2^{\frac{2m-1}{2m}}}\alpha g(z)A_1^2B_2\frac{I}{\Gamma\left(\frac{1}{2m}\right)}$$

$$+\epsilon\frac{1}{A_2B_2}\frac{m2^{\frac{2m+1}{2m}}}{\Gamma\left(\frac{1}{2m}\right)}\int_{-\infty}^{\infty}\{B_2\Im[R_2e^{-i\phi}](3-4m\tau_2^{2m})$$

$$+4\kappa_2\Re[R_2e^{-i\phi_2}]\tau_2\}e^{-\tau_2^{2m}}d\tau_2 \tag{4.105}$$

Here, once again, Eqs.(4.94)–(4.105) all respectively reduce to Eqs.(4.82)–(4.93) for the special case $m=1$. These are the parameter dynamics of chirped Gaussian and SG solitons can be very useful to study vector solitons in a birefringent media.

References

1. F. Abdullaev, S. Darmanyan & P. Khabibullaev. *Optical Solitons.* Springer-Verlag, New York, NY. USA. (1993).
2. M. J. Ablowitz & H. Segur. *Solitons and the Inverse Scattering Transform.* SIAM. Philadelphia, PA. USA. (1981).
3. M. J. Ablowitz, G. Biondini, S. Chakravarty, R. B. Jenkins & J. R. Sauer. "Four-wave mixing in wavelength-division multiplexed soliton systems—Ideal fibers. *Journal of Optical Society of America B.* Vol 14, 1788-1794. (1997).
4. G. P. Agrawal. *Nonlinear Fiber Optics.* Academic Press, San Deigo, CA. USA. (1995).

5. N. N. Akhmediev & A. Ankiewicz. *Solitons, Nonlinear Pulses and Beams.* Chapman and Hall, London. UK. (1997).
6. D. Anderson. "Variational approach to nonlinear pulse propagation in optical fibers". *Physical Review A.* Vol 27, No 6, 3135-3145. (1983).
7. A. Biswas. "Dynamics of Gaussian and super-Gaussian solitons in birefringent optical fibers". *Progress in Electromagnetics Research.* Vol 33, 119-139. (2001).
8. A. Biswas. "Dispersion-managed vector solitons in optical fibers". *Fiber and Integrated Optics.* Vol 20, No 5, 503-515. (2001).
9. A. Biswas. "Gaussian solitons in birefringent optical fibers". *International Journal of Pure and Applied Mathematics.* Vol 2, No 1, 87-104. (2002).
10. A. Biswas. "Super-Gaussian solitons in optical fibers". *Fiber and Integrated Optics.* Vol 21, No 2, 115-124. (2002).
11. A. Biswas. "Dispersion-managed solitons in optical fibers". *Journal of Optics A.* Vol 4, No 1, 84-97. (2002).
12. A. Biswas. "Integro-differential perturbation of dispersion-managed solitons". *Journal of Electromagnetic Waves and Applications.* Vol 17, No 4, 641-665. (2003).
13. A. Biswas. "Dispersion-managed solitons in optical couplers". *Journal of Nonlinear Optical Physics and Materials.* Vol 12, No 1, 45-74. (2003).
14. A. Biswas. "Perturbations of dispersion-managed optical solitons". *Progress in Electromagnetics Research.* Vol 48, 85-123. (2004).
15. A. Biswas. "Dispersion-Managed solitons in birefringent fibers and multiple channels". *International Mathematical Journal.* Vol 5, No 5, 477-515. (2004).
16. A. Biswas & S. Konar. *Introduction to Non-Kerr Law Optical Solitons.* CRC Press, Boca Raton, FL. USA. (2006).
17. R. Kohl, A. Biswas, D. Milovic & E. Zerrad. "Perturbation of Gaussian optical solitons in dispersion-managed fibers". *Applied Mathematics and Computation.* Vol 199, No 1, 250-258. (2009).
18. R. Kohl, A. Biswas, D. Milovic & E. Zerrad. "Perturbation of super-sech solitons in dispersion-managed optical fibers". *International Journal of Theoretical Physics.* Vol 47, No 7, 2038-2064. (2008).
19. R. Kohl, A. Biswas, D. Milovic & E. Zerrad. "Adiabatic dynamics of Gaussian and super-Gaussian solitons in dispersion-managed optical fibers". *Progress in Electromagnetics Research.* Vol 84, 27-53. (2008).
20. R. Kohl, D. Milovic, E. Zerrad & A. Biswas. "Perturbation of super-Gaussian optical solitons in dispersion-managed fibers". *Mathematical and Computer Modelling.* Vol 49, No 7-8, 418, 427. (2009).
21. R. Kohl, D. Milovic, E. Zerrad & A. Biswas. "Optical solitons by He's variational principle in a non-Kerr law media". *Journal of Infrared, Millimeter and Terahertz Waves.* Vol 30, No 5, 526-537. (2009).
22. R. Kohl, D. Milovic, E. Zerrad & A. Biswas. "Soliton perturbation theory for dispersion-managed optical fibers". *Journal of Nonlinear Optical Physics and Materials.* Vol 18, No 2. (2009).
23. M. F. Mahmood & S. B. Qadri. "Modeling propagation of chirped solitons in an elliptically low birefringent single-mode optical fiber". *Journal of Nonlinear Optical Physics and Materials.* Vol 8, No 4, 469-475. (1999).
24. D. Marcuse, C. R. Menyuk & P. K. A. Wai "Application of the Manakov-PMD equation to studies of signal propagation in optical fibers with randomly varying birefringence". *Journal of Lightwave Technology.* Vol 15, No 9, 1735-1746. (1997).
25. M. Matsumoto. "Analysis of interaction between stretched pulses propagating in dispersion-managed fibers". *IEEE Photonics Technology Letters.* Vol 10, No 3, 373-375. (1998).
26. A. Panajotovic, D. Milovic & A. Mittic. "Boundary case of pulse propagation analytic solution in the presence of interference and higher order dispersion". *TELSIKS 2005 Conference Proceedings.* 547-550. Nis-Serbia. (2005).

27. A. Panajotovic, D. Milovic, A. Biswas & E. Zerrad. "Influence of even order dispersion on super-sech soliton transmission quality under coherent crosstalk". *Research Letters in Optics*. Vol 2008, 613986, 5 pages. (2008).
28. A. Panajotovic, D. Milovic & A. Biswas. "Influence of even order dispersion on soliton transmission quality with coherent interference". *Progress in Electromagnetics Research B*. Vol 3, 63-72. (2008).
29. O. V. Sinkin, V. S. Grigoryan & C. R. Menyuk. "Accurate probabilistic treatment of bit-pattern-dependent nonlinear distortions in BER calculations for WDM RZ systems". *Journal of Lightwave Technology*. Vol 25, No 10, 2959-2967. (2007).
30. C. D. Stacey, R. M. Jenkins, J. Banerji & A. R. Davis. "Demonstration of fundamental mode only propagation in highly multimode fibre for high power EDFAs". *Optics Communications*. Vol 269, 310-314. (2007).
31. M. Stefanovic & D. Milovic. "The impact of out-of-band crosstalk on optical communication link preferences". *Journal of Optical Communications*. Vol 26, No 2, 69-72. (2005).
32. M. Stefanovic, D. Draca, A. Panajotovic & D. Milovic. "Individual and joint influence of second and third order dispersion on transmission quality in the presence of coherent interference". *Optik*. Vol 120, No 13, 636-641.(2009).
33. B. Stojanovic, D. M. Milovic & A. Biswas. "Timing shift of optical pulses due to interchannel crosstalk". *Progress in Electromagnetics Research M*. Vol 1, 21-30. (2008).
34. S. K. Turitsyn, I. Gabitov, E. W. Laedke, V. K. Mezentsev, S. L. Musher, E. G. Shapiro, T. Schäfer, & K. H. Spatschek. "Variational approach to optical pulse propagation in dispersion compensated transmission system". *Optics Communications*. Vol 151, No 1-3, 117-135. (1998).
35. T. R. Wolinski. "Polarimetric optical fibers and sensors". *Progress in Optics*. Vol XL, 1-75. (2000).

Chapter 5
Multiple Channels

5.1 Introduction

The successful design of low-loss dispersion-shifted and dispersion-flattened optical fibers with the low dispersion over a relatively large wavelength range can be used to reduce or completely eliminate the group velocity mismatch for the multi-channel WDM systems resulting in the desirable simultaneous arrival of time aligned bit pulses, thus creating a new class of bit-parallel wavelength links that is used in high-speed single fiber computer buses. In spite of the intrinsically small value of the nonlinearity-induced change in the refractive index of the fused silica, nonlinear effects in optical fibers cannot be ignored even at relatively low powers. In particular, in WDM systems with the simultaneous transmission of pulses of different wavelengths, the cross-phase modulation (XPM) effects needs to be taken into account. Although the XPM will not cause the energy to be exchanged among the different wavelengths, it will lead to the interaction of pulses and thus the pulse positions and shapes gets altered significantly. The multi-channel WDM transmission of co-propagating wave envelopes in a nonlinear optical fiber, including the XPM effect, can be modeled by the following N-coupled NLSE in the dimensionless form [5]:

$$iq_z^{(l)} + \frac{D(z)}{2} q_{tt}^{(l)} + g(z) \left\{ \left| q^{(l)} \right|^2 + \sum_{m \neq l}^{N} \alpha_{lm} \left| q^{(m)} \right|^2 \right\} q^{(l)} = 0 \tag{5.1}$$

where $1 \leq l \leq N$. Equation (5.1) is the model for the bit-parallel WDM soliton transmission. Here α_{lm} are known as the XPM coefficients. It is well known [32] that the straightforward use of this system for the description of WDM transmission could potentially give incorrect results. However, this model can be applied to describe the WDM transmission for dispersion flattened fibers, the dispersion of which weakly depends on the operating wavelength.

It was first pointed out by Yeh and Bergman [5] in model (5.1) that when two or more optical pulses co-propagate simultaneously and affect each other through the intensity dependence of the refractive index, the XPM term can be used to produce an interesting *pulse shepherding effect.* In particular, Yeh and Bergman studied (5.1) numerically the evolution of two pulses whose operating wavelengths are separated by 4 nm ($\lambda_1 = 1.55$ μm and $\lambda_2 = 1.546$ μm). For this case, the FWM effect is negligible. When these pulses are initially offset by a half pulse width, the pulses tend to attract each other. However, when the two co-propagating pulses, on two separate wavelengths, are separated by a sufficiently large distance, these two pulses will not interact with each other. Two widely separated pulses can be brought sufficiently close to each other by launching another pulse on a separate wavelength (λ_3 $= 1.542$ μm) with the proper magnitude and at the proper time. According to Yeh and Bergman this pulse is called the *shepherd pulse* because of its shepherding behavior with other pulses. Computer experiments showed that a low-magnitude shepherd pulse does not posses sufficient attractive strength to pull the shepherd pulses together. Additionally, it was noticed that at a very large amplitude of the shepherd pulse, the shepherded pulses tend to break up, and for the broadened shepherd pulse the effect is getting weaker. Therefore, these experiments suggest that, for Eq.(5.1), there exists an optimum pulse with a certain magnitude, pulse width and pulse shape that can provide the best alignment for these pulses. As a matter of fact, this observation means that the most effective shepherding effect should be observed for pulses that are close to the so called "nonlinear modes" of the model, that is solitons of the nonlinear model (5.1).

Another important medium in which the model given by Eq.(5.1) arises is the photo-refractive medium [5]. In the case of incoherent beam propagation in a biased photo-refractive crystal, which is a non-instantaneous nonlinear media, the diffraction behavior of that incoherent beam is to be treated somewhat differently. The diffraction behavior of an incoherent beam can be effectively described by the sum of the intensity contributions from all its coherent components. Then the governing equation of N self-trapped mutually incoherent wave packets in such a media is given by Eq.(5.1).

Equation (5.1) is, in general, not integrable. However, it can be solved analytically for certain very specific cases, namely, when $D(z) = g(z) = 1$ along with $\alpha_{lm} = 1$, $\forall m, n$. In Eq.(5.1), the special case, where $N = 2$, reduces to the case of birefringent fibers. This case is studied in the previous chapter.

5.2 Integrals of motion

It needs to be noted that Eq. (5.1) does not have infinitely many conservation laws. In fact, it has at least two integrals of motion and they are the energy

(E) and the linear momentum (M) that are respectively given by [5]

$$E = \sum_{l=1}^{N} \int_{-\infty}^{\infty} \left| q^{(l)} \right|^2 dt \tag{5.2}$$

and

$$M = \frac{i}{2} D(z) \sum_{l=1}^{N} \int_{-\infty}^{\infty} \left(q^{(l)*} q_t^{(l)} - q^{(l)} q_t^{(l)*} \right) dt \tag{5.3}$$

The Hamiltonian (H) given by

$$H = \frac{1}{2} \sum_{l=1}^{N} \int_{-\infty}^{\infty} \left\{ D(z) \left| q_t^{(l)} \right|^2 - g(z) \sum_{m=1}^{N} \alpha_{lm} \left| q^{(l)} \right|^2 \left| q^{(m)} \right|^2 \right\} dt \tag{5.4}$$

is, however, not a conserved quantity unless, in addition to $D(z)$ and $g(z)$ being constants, the matrix of XPM coefficients $\Lambda = (\alpha_{ij})_{N \times N}$ is a symmetric matrix, namely, $\alpha_{ij} = \alpha_{ji}$ for $1 \leq i, j \leq N$. Thus, for a birefringent fiber, the matrix should be of the form

$$\Lambda = \begin{bmatrix} 0 & \alpha_{12} \\ \alpha_{12} & 0 \end{bmatrix} \tag{5.5}$$

while for a triple channeled fiber,

$$\Lambda = \begin{bmatrix} 0 & \alpha_{12} & \alpha_{13} \\ \alpha_{12} & 0 & \alpha_{23} \\ \alpha_{13} & \alpha_{23} & 0 \end{bmatrix} \tag{5.6}$$

and so on. Now, it is assumed that the solution of Eq.(5.1) is given by a chirped pulse, in the lth core, of the form [5]

$$\begin{aligned} q^{(l)}(z,t) = A_l(z) f\left[B_l(z) \left\{ t - \bar{t}_l(z) \right\} \right] \exp[& iC_l(z) \left\{ t - \bar{t}_l(z) \right\}^2 \\ & - i\kappa_l(z) \left\{ t - \bar{t}_l(z) \right\} + i\theta_l(z)] \end{aligned} \tag{5.7}$$

where f represents the shape of the pulse. It could be a Gaussian type or an SG type pulse. Also, here the parameters $A_l(z)$, $B_l(z)$, $C_l(z)$, $\kappa_l(z)$, $\bar{t}_l(z)$ and $\theta_l(z)$ respectively represent the soliton amplitude, the inverse width of the pulse, chirp, frequency, the center of the pulse and the phase of the pulse in the lth channel. Using the variational principle, a set of evolution equations for the pulse parameters will be derived. Once again, this approach is only approximate and does not account for characteristics such as the energy loss due to the continuum radiation, damping of the amplitude oscillations and changing of the pulse shape. For convenience, the following integrals are defined:

$$I_{a,b,c}^{(l)} = \int_{-\infty}^{\infty} \tau^a f^b(\tau) \left(\frac{df}{d\tau}\right)^c d\tau \tag{5.8}$$

$$J_{l,m} = \int_{-\infty}^{\infty} \prod_{j=l,m} f^2 \left[B_j(z)\left(t - \bar{t}_j(z)\right)\right] dt \tag{5.9}$$

where a, b and c are integers and $1 \leq l \leq N$. For such a pulse form given by (5.7), integrals of motion are

$$E = \sum_{l=1}^{N} \int_{-\infty}^{\infty} \left|q^{(l)}\right|^2 dt = \sum_{l=1}^{N} \frac{A_l^2}{B_l} I_{0,2,0}^{(l)} \tag{5.10}$$

$$\begin{aligned} M &= \frac{i}{2} D(z) \sum_{l=1}^{N} \int_{-\infty}^{\infty} \left(q^{(l)} q_t^{(l)*} - q^{(l)*} q_t^{(l)}\right) dt \\ &= \sum_{l=1}^{N} \kappa_l \frac{A_l^2}{B_l} I_{0,2,0}^{(l)} \end{aligned} \tag{5.11}$$

while the Hamiltonian is

$$\begin{aligned} H &= \frac{1}{2} \sum_{l=1}^{N} \int_{-\infty}^{\infty} \left\{ D(z) \left|q_t^{(l)}\right|^2 - g(z) \sum_{m=1}^{N} \alpha_{lm} \left|q^{(l)}\right|^2 \left|q^{(m)}\right|^2 \right\} dt \\ &= \frac{D(z)}{2} \sum_{l=1}^{N} \left(A_l^2 B_l I_{0,0,2}^{(l)} + 4 \frac{A_l^2 C_l^2}{B_l^3} I_{2,2,0}^{(l)} + \frac{\kappa_l^2 A_l^2}{B_l} I_{0,2,0}^{(l)} \right) \\ &\quad - \frac{g(z)}{2} \sum_{l=1}^{N} \sum_{m \neq l}^{N} \alpha_{lm} A_l^2 A_m^2 J_{l,m} \end{aligned} \tag{5.12}$$

5.2.1 Gaussian pulses

For a pulse of Gaussian type, substitute $f(\tau) = e^{-\frac{1}{2}\tau^2}$. Thus, the conserved quantities respectively reduce to

$$E = \sum_{l=1}^{N} \int_{-\infty}^{\infty} \left|q^{(l)}\right|^2 dt = \sqrt{\frac{\pi}{2}} \sum_{l=1}^{N} \frac{A_l^2}{B_l} \tag{5.13}$$

$$M = \frac{i}{2} D(z) \sum_{l=1}^{N} \int_{-\infty}^{\infty} \left(q^{(l)} q_t^{(l)*} - q^{(l)*} q_t^{(l)}\right) dt$$

$$= D(z)\sqrt{\frac{\pi}{2}}\sum_{l=1}^{N}\kappa_l\frac{A_l^2}{B_l} \tag{5.14}$$

while the Hamiltonian is

$$H = \frac{1}{2}\sum_{l=1}^{N}\int_{-\infty}^{\infty}\left\{D(z)\left|q_t^{(l)}\right|^2 - g(z)\sum_{m=1}^{N}\alpha_{lm}\left|q^{(l)}\right|^2\left|q^{(m)}\right|^2\right\}dt$$

$$= \frac{D(z)}{2}\sqrt{\frac{\pi}{2}}\sum_{l=1}^{N}\left(A_l^2B_l + \frac{A_l^2C_l^2}{B_l^3} + \frac{\kappa_l^2A_l^2}{B_l}\right)$$

$$-\frac{g(z)}{2}\sum_{l=1}^{N}\sum_{m\neq l}^{N}\alpha_{lm}A_l^2A_m^2\sqrt{\frac{2\pi}{B_l^2+B_m^2}}$$

$$\cdot\exp\left\{\frac{B_l^2B_m^2}{2\left(B_l^2+B_m^2\right)}\left(\bar{t}_l-\bar{t}_m\right)^2\right\} \tag{5.15}$$

where, for Gaussian pulses, the integral $J_{l,m}$, from Eq.(5.9) reduces to

$$J_{l,m} = \sqrt{\frac{2\pi}{B_l^2+B_m^2}}\exp\left\{\frac{B_l^2B_m^2}{2\left(B_l^2+B_m^2\right)}\left(\bar{t}_l-\bar{t}_m\right)^2\right\} \tag{5.16}$$

5.2.2 Super-Gaussian pulses

For SG pulse, choose $f(\tau) = e^{-\frac{1}{2}\tau^{2p}}$ with $p \geq 1$, where the parameter p controls the degree of edge sharpness. For an SG pulse, the integral $J_{l,m}$, from Eq.(5.9) reduces to

$$J_{l,m} = \int_{-\infty}^{\infty}\exp\left[-\frac{1}{2}\left\{B_l^{2p}\left(t-\bar{t}_l\right)^{2p} + B_m^{2p}\left(t-\bar{t}_m\right)^{2p}\right\}\right] \tag{5.17}$$

which cannot be obtained in a closed form, unlike in the case of Gaussian soliton, and unless a particular value of p is considered. So, the integrals of motion respectively are

$$E = \sum_{l=1}^{N}\int_{-\infty}^{\infty}\left|q^{(l)}\right|^2dt = \frac{1}{p2^{\frac{1}{2p}}}\Gamma\left(\frac{1}{2p}\right)\sum_{l=1}^{N}\frac{A_l^2}{B_l} \tag{5.18}$$

$$M = \frac{i}{2}D(z)\sum_{l=1}^{N}\int_{-\infty}^{\infty}\left(q^{(l)}q_t^{(l)*} - q^{(l)*}q_t^{(l)}\right)dt$$

$$= \frac{D(z)}{p2^{\frac{1}{2p}}}\Gamma\left(\frac{1}{2p}\right)\sum_{l=1}^{N}\kappa_l\frac{A_l^2}{B_l} \tag{5.19}$$

while the Hamiltonian is

$$H = \frac{1}{2}\sum_{l=1}^{N}\int_{-\infty}^{\infty}\left\{D(z)\left|q_t^{(l)}\right|^2 - g(z)\sum_{m=1}^{N}\alpha_{lm}\left|q^{(l)}\right|^2\left|q^{(m)}\right|^2\right\}dt$$

$$= \sum_{l=1}^{N}\left[\frac{D(z)}{2}\left\{\frac{A_l^2B_l}{2}\Gamma\left(\frac{2p-1}{2p}\right) + 4\frac{A_l^2C_l^2}{pB_l^3}\Gamma\left(\frac{3}{2p}\right)\right.\right.$$

$$\left.\left. + \frac{\kappa_l^2A_l^2}{pB_l}\Gamma\left(\frac{1}{2p}\right)\right\} - \frac{g(z)}{2}\sum_{m\neq l}^{N}\alpha_{lm}A_l^2A_m^2J_{l,m}\right] \tag{5.20}$$

5.3 Variational principle

The soliton parameter dynamics will be obtained by the aid of variational principle. In order to study this principle, the Lagrangian of Eq.(5.1) is needed. This is given by [5]

$$L = \frac{1}{2}\sum_{l=1}^{N}\int_{-\infty}^{\infty}\left[i\left(q^{(l)*}q_z^{(l)} - q^{(l)}q_z^{(l)*}\right) - D(z)\left|q_t^{(l)}\right|^2\right.$$

$$\left. + g(z)\left|q^{(l)}\right|^4 + 2g(z)\sum_{m=1}^{N}\alpha_{lm}\left|q^{(l)}\right|^2\left|q^{(m)}\right|^2\right]dt \tag{5.21}$$

Now, using Eqs.(5.7)–(5.9), the Lagrangian given by Eq.(5.21), reduces to

$$L = \sum_{l=1}^{N}\left[-D(z)\left\{\frac{A_l^2B_l}{2}I_{0,0,2}^{(l)} + 2\frac{A_l^2C_l^2}{B_l^3}I_{2,2,0}^{(l)} + \frac{\kappa_l^2A_l^2}{2B_l}I_{0,2,0}^{(l)}\right\}\right.$$

$$- \frac{g(z)}{2}\frac{A_l^4}{B_l}I_{0,4,0}^{(l)} - \frac{A_l^2}{B_l^3}\frac{dC_l}{dz}I_{2,2,0}^{(l)} + \frac{A_l^2}{B_l}\left(\bar{t}_l\frac{d\kappa_l}{dz} - \frac{d\theta_l}{dz}\right)I_{0,2,0}^{(l)}$$

$$\left. + g(z)A_l^2\sum_{m\neq l}^{N}\alpha_{lm}A_m^2J_{l,m}\right] \tag{5.22}$$

By the EL equation, seen in Chapter 2, the soliton parameter dynamics are

$$\frac{dA_l}{dz} = -D(z)A_lC_l \tag{5.23}$$

$$\frac{dB_l}{dz} = -2D(z)B_lC_l \tag{5.24}$$

$$\frac{dC_l}{dz} = \frac{D(z)}{2}\left\{B_l^4\frac{I_{0,0,2}^{(l)}}{I_{2,2,0}^{(l)}} - 4C_l^2\right\}$$

$$-\frac{g(z)}{4}A_l^2B_l^2\frac{I_{0,4,0}^{(l)}}{I_{2,2,0}^{(l)}} - \frac{g(z)}{2I_{2,2,0}^{(l)}}B_l^3\sum_{m\neq l}^{N}\alpha_{lm}A_m^2J_{l,m} \tag{5.25}$$

$$\frac{d\kappa_l}{dz} = 0 \tag{5.26}$$

$$\frac{d\bar{t}_l}{dz} = -D(z)\kappa_l \tag{5.27}$$

$$\frac{d\theta_l}{dz} = \frac{D(z)}{2}\left\{\kappa_l^2 - 2B_l^2\frac{I_{0,0,2}^{(l)}}{I_{0,2,0}^{(l)}}\right\} + \frac{5}{4}g(z)A_l^2\frac{I_{0,4,0}^{(l)}}{I_{0,2,0}^{(l)}}$$

$$+\frac{3}{2}\frac{g(z)}{I_{0,2,0}^{(l)}}B_l\sum_{m\neq l}^{N}\alpha_{lm}A_m^2J_{l,m} \tag{5.28}$$

This dynamical system of soliton parameters will now be modified to the two special cases where two types of pulses will be considered. They are the Gaussian and super-Gaussian pulses that are studied in the following two subsections.

5.3.1 Gaussian pulses

For Gaussian pulses, the pulse parameter dynamics reduces to

$$\frac{dA_l}{dz} = -D(z)A_lC_l \tag{5.29}$$

$$\frac{dB_l}{dz} = -2D(z)B_lC_l \tag{5.30}$$

$$\frac{dC_l}{dz} = 2D(z)\left(B_l^4 - C_l^2\right) - \frac{\sqrt{2}}{2}g(z)A_l^4B_l - \sqrt{2}g(z)B_l^3$$

$$\cdot\sum_{m\neq l}^{N}\frac{\alpha_{lm}A_m^4}{B_m\sqrt{B_l^2+B_m^2}}\exp\left\{\frac{B_l^2B_m^2}{2\left(B_l^2+B_m^2\right)}\left(\bar{t}_l-\bar{t}_m\right)^2\right\} \tag{5.31}$$

$$\frac{d\kappa_l}{dz} = 0 \tag{5.32}$$

$$\frac{d\bar{t}_l}{dz} = -D(z)\kappa_l \tag{5.33}$$

$$\frac{d\theta_l}{dz} = \frac{D(z)}{2}\left(\kappa_l^2 - B_l^2\right) + \frac{5\sqrt{2}}{8}\frac{g(z)A_l^4}{B_l} + \frac{3\sqrt{2}}{2}g(z)B_l$$
$$\cdot \sum_{m\neq l}^{N} \frac{\alpha_{lm}A_m^4}{B_m\sqrt{B_l^2+B_m^2}} \exp\left\{\frac{B_l^2 B_m^2}{2\left(B_l^2+B_m^2\right)}\left(\bar{t}_l - \bar{t}_m\right)^2\right\} \tag{5.34}$$

5.3.2 Super-Gaussian pulses

For SG pulses, the dynamical system simplifies to

$$\frac{dA_l}{dz} = -D(z)A_lC_l \tag{5.35}$$

$$\frac{dB_l}{dz} = -2D(z)B_lC_l \tag{5.36}$$

$$\frac{dC_l}{dz} = \frac{D(z)}{8}\left\{B_l^4 p(2p-1)\frac{\Gamma\left(\frac{2p-1}{2p}\right)}{\Gamma\left(\frac{3}{2p}\right)} - 8C_l^2\right\}$$
$$-\frac{g(z)}{2^{\frac{4p+1}{2p}}}A_l^4 B_l\frac{\Gamma\left(\frac{1}{2p}\right)}{\Gamma\left(\frac{3}{2p}\right)} - \frac{pg(z)}{\Gamma\left(\frac{3}{2p}\right)}B_l^3\sum_{m\neq l}^{N}\frac{\alpha_{lm}A_m^4 J_{l,m}}{B_m} \tag{5.37}$$

$$\frac{d\kappa_l}{dz} = 0 \tag{5.38}$$

$$\frac{d\bar{t}_l}{dz} = -D(z)\kappa_l \tag{5.39}$$

$$\frac{d\theta_l}{dz} = \frac{D(z)}{2}\left\{\kappa_l^2 - B_l^2 p(2p-1)\frac{\Gamma\left(\frac{2p-1}{2p}\right)}{\Gamma\left(\frac{1}{2p}\right)}\right\} + \frac{5}{2^{\frac{4p+1}{2p}}}\frac{g(z)A_l^4}{B_l}$$
$$+\frac{3}{2}\frac{p}{\Gamma\left(\frac{1}{2p}\right)}g(z)B_l\sum_{m\neq l}^{N}\frac{\alpha_{lm}A_m^4 J_{l,m}}{B_m} \tag{5.40}$$

5.4 Perturbation terms

In this section, the NLSE for multiple channels in presence of perturbation terms will be studied. The perturbed DWDM system given by

$$iq_z^{(l)} + \frac{D(z)}{2} q_{tt}^{(l)} + g(z) \left\{ \left| q^{(l)} \right|^2 + \sum_{m \neq l}^{N} \alpha_{lm} \left| q^{(m)} \right|^2 \right\} q^{(l)}$$

$$= i\epsilon R \left[q^{(l)}, q^{(l)*}; q^{(m)}, q^{(m)*} \right] \tag{5.41}$$

where, once again, $1 \leq l \leq N$ while $m \neq l$. In presence of perturbation terms, the EL equation modify to [5]

$$\frac{\partial L}{\partial p} - \frac{d}{dz} \left(\frac{\partial L}{\partial p_z} \right) = i\epsilon \int_{-\infty}^{\infty} \left(R \frac{\partial q^{(l)*}}{\partial p} - R^* \frac{\partial q^{(l)}}{\partial p} \right) dt \tag{5.42}$$

where p represents the $6N$ soliton parameters. Once again, substituting A_l, B_l, C_l, κ_l, $\bar{t}_l$ and θ_l for p in Eq.(5.42), it is possible to arrive at the following adiabatic evolution equations:

$$\frac{dA_l}{dz} = -D(z) A_l C_l$$

$$+ \frac{\epsilon B_l}{4} \int_{-\infty}^{\infty} \left(\frac{\tau_l^2}{I_{2,2,0}^{(l)}} - \frac{3}{I_{0,2,0}^{(l)}} \right) (q^{(l)*} R + q^{(l)} R^*) dt \tag{5.43}$$

$$\frac{dB_l}{dz} = -2D(z) B_l C_l$$

$$+ \frac{\epsilon B_l^2}{2A_l} \int_{-\infty}^{\infty} \left(\frac{\tau_l^2}{I_{2,2,0}^{(l)}} - \frac{1}{I_{0,2,0}^{(l)}} \right) (q^{(l)*} R + q^{(l)} R^*) dt \tag{5.44}$$

$$\frac{dC_l}{dz} = \frac{D(z)}{2} \left(B_l \frac{I_{0,0,2}^{(l)}}{I_{2,2,0}^{(l)}} - 4C_l^2 \right) - \frac{g(z)}{4} A_l^2 B_l^2 \frac{I_{0,4,0}^{(l)}}{I_{2,2,0}^{(l)}}$$

$$- \frac{g(z)}{2I_{2,2,0}^{(l)}} B_l^3 \sum_{m \neq l}^{N} \alpha_{lm} A_m^2 J_{l,m}$$

$$- \frac{i\epsilon}{4} \frac{1}{A_l} \frac{1}{I_{2,2,0}^{(l)}} \int_{-\infty}^{\infty} [B_l (q^{(l)} R^* - q^{(l)*} R)$$

$$+ 2\tau_l (q_t^{(l)} R^* - q_t^{(l)*} R)] dt \tag{5.45}$$

$$\frac{d\kappa_l}{dz} = \frac{\epsilon}{A_l} \frac{1}{I_{0,2,0}^{(l)}} \int_{-\infty}^{\infty} [iB_l (q_t^{(l)} R^* - q_t^{(l)*} R)$$

$$- 2\tau_l C_l (q^{(l)*} R + q^{(l)} R^*)] dt \tag{5.46}$$

$$\frac{d\bar{t}_l}{dz} = -D(z)\kappa_l + \frac{\epsilon}{A_l} \frac{1}{I_{0,2,0}^{(l)}} \int_{-\infty}^{\infty} \tau_l \left(q^{(l)*} R + q^{(l)} R^* \right) dt \tag{5.47}$$

$$\frac{d\theta_l}{dz} = D(z)\left(\frac{\kappa_l^2}{2} - \frac{I_{0,0,2}^{(l)}}{I_{0,2,0}^{(l)}}B_l^2\right) + \frac{5g(z)}{4}A_l^2\frac{I_{0,4,0}^{(l)}}{I_{0,2,0}^{(l)}} + \frac{3}{2}\frac{g(z)}{I_{0,2,0}^{(l)}}B_l\sum_{m\neq l}^{N}\alpha_{lm}A_m^2 J_{l,m} + \frac{\epsilon}{2}\frac{1}{A_l}\frac{1}{I_{0,2,0}^{(l)}}\int_{-\infty}^{\infty}[3iB_l(q^{(l)}R^* - q^{(l)*}R) + 2i\tau_l(q_t^{(l)}R^* - q_t^{(l)*}R) + 4\kappa_l\tau_l(q^{(l)*}R + q^{(l)}R^*)]dt \tag{5.48}$$

Now, relations (5.43)–(5.48) can be rewritten in the following alternative convenient forms. These are also known as the phase-amplitude forms.

$$\frac{dA_l}{dz} = -D(z)A_lC_l - \frac{\epsilon}{2}\int_{-\infty}^{\infty}\Re[Re^{-i\phi_l}]\left(\frac{\tau_l^2}{I_{0,2,0}^{(l)}} - \frac{3}{I_{2,2,0}^{(l)}}\right)f(\tau_l)d\tau_l \tag{5.49}$$

$$\frac{dB_l}{dz} = -2D(z)B_lC_l - \epsilon\frac{B_l}{A_l}\int_{-\infty}^{\infty}\Re[Re^{-i\phi_l}]\left(\frac{\tau_l^2}{I_{2,2,0}^{(l)}} - \frac{1}{I_{0,2,0}^{(l)}}\right)f(\tau_l)d\tau_l \tag{5.50}$$

$$\frac{dC_l}{dz} = \frac{D(z)}{2}\left(B_l\frac{I_{0,0,2}^{(l)}}{I_{2,2,0}^{(l)}} - 4C_l^2\right) - \frac{g(z)}{4}A_l^2B_l^2\frac{I_{0,4,0}^{(l)}}{I_{2,2,0}^{(l)}} - \frac{g(z)}{2I_{2,2,0}^{(l)}}B_l^3\sum_{m\neq l}^{N}\alpha_{lm}A_m^2J_{l,m} - \frac{\epsilon B_l^2}{2A_lI_{2,2,0}^{(l)}}\int_{-\infty}^{\infty}\Im[Re^{-i\phi_l}]\left(f(\tau_l) + 2\tau_l\frac{df}{d\tau_l}\right)d\tau_l \tag{5.51}$$

$$\frac{d\kappa_l}{dz} = \frac{2\epsilon}{A_lB_lI_{0,2,0}^{(l)}}\int_{-\infty}^{\infty}\left\{B_l^2\Im[Re^{-i\phi_l}]\frac{df}{d\tau_l} - 2C_l\Re[Re^{-i\phi_l}]\tau f(\tau_l)\right\}d\tau_l \tag{5.52}$$

$$\frac{d\bar{t}_l}{dz} = -D(z)\kappa_l + \frac{2\epsilon}{A_lB_lI_{0,2,0}^{(l)}}\int_{-\infty}^{\infty}\Re[Re^{-i\phi_l}]\tau f(\tau_l)d\tau_l \tag{5.53}$$

$$\frac{d\theta_l}{dz} = \frac{D(z)}{2}\left\{\kappa_l^2 - 2B_l^2\frac{I_{0,0,2}^{(l)}}{I_{0,2,0}^{(l)}}\right\} + \frac{5}{4}g(z)A_l^2\frac{I_{0,4,0}^{(l)}}{I_{0,2,0}^{(l)}} + \frac{3}{2}\frac{g(z)}{I_{0,2,0}^{(l)}}B_l\sum_{m\neq l}^{N}\alpha_{lm}A_m^2J_{l,m} + \frac{\epsilon}{2A_lB_lI_{0,2,0}^{(l)}}\int_{-\infty}^{\infty}\left\{B_l\Im[Re^{-i\phi_l}]\cdot\left(3f(\tau_l) + 2\tau_l\frac{df}{d\tau_l}\right) + 4\kappa\Re[Re^{-i\phi_l}]\tau f(\tau_l)\right\}d\tau_l \tag{5.54}$$

where

$$\tau_l = B(z)\left(t - \bar{t}_l(z)\right) \tag{5.55}$$

and

$$\phi_l = C_l(z)\left\{t - \bar{t}_l(z)\right\}^2 - \kappa_l(z)\left\{t - \bar{t}_l(z)\right\} + \theta_l(z) \tag{5.56}$$

for $1 \leq l \leq N$. Also, $\Re$ and $\Im$ represent the real and imaginary parts, respectively.

5.4.1 Gaussian pulses

Now, substituting the integrals $I^{(l)}_{a,b,c}$ for the indicated values of a, b and c in Eqs.(5.49)–(5.54), the adiabatic parameter dynamics for Gaussian pulse are

$$\frac{dA_l}{dz} = -D(z)A_lC_l - \frac{\epsilon}{\sqrt{2\pi}}\int_{-\infty}^{\infty} \Re[Re^{-i\phi_l}](4\tau_l^2 - 3)e^{-\frac{1}{2}\tau_l^2}\,d\tau_l \tag{5.57}$$

$$\frac{dB_l}{dz} = -2D(z)B_lC_l - \epsilon\sqrt{\frac{2}{\pi}\frac{B_l}{A_l}}\int_{-\infty}^{\infty} \Re[Re^{-i\phi_l}](4\tau_l^2 - 1)e^{-\frac{1}{2}\tau_l^2}\,d\tau_l \tag{5.58}$$

$$\begin{aligned}\frac{dC_l}{dz} &= 2D(z)(B_l^4 - C_l^2) - \frac{\sqrt{2}}{2}g(z)A_l^4B_l \\ &\quad - \sqrt{2}g(z)B_l^3\sum_{m\neq l}^{N}\frac{\alpha_{lm}A_m^4}{B_m\sqrt{B_l^2 + B_m^2}} \\ &\quad \cdot \exp\left\{\frac{B_l^2B_m^2}{2\left(B_l^2 + B_m^2\right)}\left(\bar{t}_l - \bar{t}_m\right)^2\right\} \\ &\quad - 2\epsilon\sqrt{\frac{2}{\pi}\frac{B_l^2}{A_l}}\int_{-\infty}^{\infty} \Im[Re^{-i\phi_l}](1 - 4\tau_l^2)e^{-\frac{1}{2}\tau_l^2}\,d\tau_l\end{aligned} \tag{5.59}$$

$$\frac{d\kappa_l}{dz} = -\epsilon\sqrt{\frac{2}{\pi}}\frac{4}{A_lB_l}\int_{-\infty}^{\infty}\left\{\tau_lB_l^2\Im[Re^{-i\phi_l}] + \tau_lC_l\Re[Re^{-i\phi_l}]\right\}e^{-\frac{1}{2}\tau_l^2}\,d\tau_l \tag{5.60}$$

$$\frac{d\bar{t}_l}{dz} = -D(z)\kappa_l + \epsilon\frac{\sqrt{2\pi}}{A_lB_l}\int_{-\infty}^{\infty} \Re[Re^{-i\phi_l}]\tau_le^{-\frac{1}{2}\tau_l^2}\,d\tau_l \tag{5.61}$$

$$\begin{aligned}\frac{d\theta_l}{dz} &= \frac{D(z)}{2}(\kappa_l^2 - B_l^2) + \frac{5\sqrt{2}}{8}g(z)\frac{A_l^4}{B_l} + \frac{3\sqrt{2}}{2}g(z)B_l \\ &\quad \cdot \sum_{m\neq l}^{N}\frac{\alpha_{lm}A_m^4}{B_m\sqrt{B_l^2 + B_m^2}}\exp\left\{\frac{B_l^2B_m^2}{2\left(B_l^2 + B_m^2\right)}\left(\bar{t}_l - \bar{t}_m\right)^2\right\}\end{aligned}$$

$$+\frac{\epsilon}{\sqrt{2\pi}}\frac{1}{A_lB_l}\int_{-\infty}^{\infty}\{B_l\Im[Re^{-i\phi_l}](3-4\tau_l^2)$$
$$+4\tau_l\kappa_l\Re[Re^{-i\phi_l}]\tau_l\}e^{-\frac{1}{2}\tau^2}d\tau_l \tag{5.62}$$

These equations now represent the evolution equations for the parameters of a Gaussian pulse, for $1 \le l \le N$, propagating through an optical fiber in presence of the perturbation terms.

5.4.2 Super-Gaussian pulses

For the perturbation terms of an SG pulse, substituting the integrals $I_{a,b,c}^{(l)}$ for $1 \le l \le N$ and the form of $f(\tau_l)$ in Eqs.(5.49)–(5.54) leads to

$$\frac{dA_l}{dz}=-D(z)A_lC_l-\epsilon p2^{\frac{p+2}{p}}\int_{-\infty}^{\infty}\Re[Re^{-i\phi_l}]$$
$$\cdot\left\{\frac{\tau_l^2}{2^{\frac{1}{p}}}\frac{1}{\Gamma(\frac{3}{2p})}-\frac{3}{2^{\frac{3}{2p}}}\frac{1}{\Gamma(\frac{1}{2p})}\right\}e^{-\frac{1}{2}\tau_l^{2p}}d\tau_l \tag{5.63}$$

$$\frac{dB_l}{dz}=-2D(z)B_lC_l-\epsilon\frac{B_l}{A_l}p2^{\frac{2}{p}}\int_{-\infty}^{\infty}\Re[Re^{-i\phi_l}]$$
$$\cdot\left\{\frac{\tau_l^2}{2^{\frac{1}{2p}}}\frac{1}{\Gamma(\frac{3}{2p})}-\frac{1}{2^{\frac{3}{2p}}}\frac{1}{\Gamma(\frac{1}{2p})}\right\}e^{-\frac{1}{2}\tau_l^{2p}}d\tau_l \tag{5.64}$$

$$\frac{dC_l}{dz}=\frac{D(z)}{8}\left\{B_l^4p(2p-1)\frac{\Gamma(\frac{2p-1}{2p})}{\Gamma(\frac{3}{2p})}-8C_l^2\right\}$$
$$-\frac{g(z)}{2^{\frac{4p+1}{2p}}}A_l^4B_l\frac{\Gamma(\frac{1}{2p})}{\Gamma(\frac{3}{2p})}-\frac{pg(z)}{\Gamma(\frac{3}{2p})}B_l^3\sum_{m\neq l}^{N}\alpha_{lm}\frac{A_m^4}{B_m}J_{l,m}$$
$$-\epsilon\frac{B_l^2}{A_l}\frac{p2^{\frac{2p+3}{2p}}}{\Gamma(\frac{3}{2p})}\int_{-\infty}^{\infty}\Im[Re^{-i\phi_l}](1-4p\tau_l^{2p})e^{-\frac{1}{2}\tau_l^{2p}}d\tau_l \tag{5.65}$$

$$\frac{d\kappa_l}{dz}=-\epsilon\frac{1}{A_lB_l}\frac{p2^{\frac{2p-1}{2p}}}{\Gamma(\frac{1}{2p})}\int_{-\infty}^{\infty}\{2p\tau_l^{2p-1}B_l^2\Im[Re^{-i\phi_l}]$$
$$+2\tau_lC_l\Re[Re^{-i\phi_l}]\}e^{-\frac{1}{2}\tau_l^{2p}}d\tau_l \tag{5.66}$$

$$\frac{d\bar{t}_l}{dz}=-D(z)\kappa_l+\epsilon\frac{1}{A_lB_l}\frac{p2^{\frac{2p+1}{2p}}}{\Gamma(\frac{1}{2p})}\int_{-\infty}^{\infty}\Re[Re^{-i\phi_l}]\tau_le^{-\frac{1}{2}\tau_l^{2p}}d\tau_l \tag{5.67}$$

$$\frac{d\theta_l}{dz} = \frac{D(z)}{2}\left\{\kappa_l^2 - B_l^2 p(2p-1)\frac{\Gamma\left(\frac{2p-1}{2p}\right)}{\Gamma\left(\frac{1}{2p}\right)}\right\}$$
$$+ \frac{5}{2^{\frac{4p+1}{2p}}} g(z)\frac{A_l^4}{B_l} + \frac{3}{2}\frac{p}{\Gamma\left(\frac{1}{2p}\right)} g(z) B_l \sum_{m\neq l}^{N} \alpha_{lm}\frac{A_m^4}{B_m} J_{l,m}$$
$$+ \epsilon\frac{1}{A_l B_l}\frac{p2^{\frac{2p+1}{2p}}}{\Gamma\left(\frac{1}{2p}\right)}\int_{-\infty}^{\infty}\{B_l \Im[Re^{-i\phi_l}](3 - 4p\tau_l^{2p})$$
$$+ 4\kappa_l \Re[Re^{-i\phi_l}]\tau_l\} e^{-\frac{1}{2}\tau_l^{2p}} d\tau_l \tag{5.68}$$

So, now, these are the adiabatic evolution of the soliton parameters for an SG pulse, for $1 \leq l \leq N$, in presence of the perturbation terms.

References

1. M. J. Ablowitz & G. Biondini. "Multiscale pulse dynamics in communication systems with strong dispersion management". *Optics Letters.* Vol 23, No 21, 1668-1670. (1998).
2. G. P. Agrawal. *Nonlinear Fiber Optics.* Academic Press, San Deigo, CA. USA. (1995).
3. N. N. Akhmediev & A. Ankiewicz. *Solitons, Nonlinear Pulses and Beams.* Chapman and Hall, London. UK. (1997).
4. D. Anderson. "Variational approach to nonlinear pulse propagation in optical fibers". *Physical Review A.* Vol 27, No 6, 3135-3145. (1983).
5. A. Biswas. "Dispersion-Managed solitons in multiple channels". *Journal of Nonlinear Optical Physics and Materials.* Vol 13, No 1, 81-102. (2004).
6. T. Hirooka & A. Hasegawa. "Chirped soliton interaction in strongly dispersion-managed wavelength-division-multiplexing systems". *Optics Letters.* Vol 23, No 10, 768-770. (1998).
7. T. Hirooka & S. Wabnitz. "Stabilization of dispersion-managed soliton transmission by nonlinear gain". *Electronics Letters.* Vol 35, No 8, 655-657. (1999).
8. T. Hirooka & S. Wabnitz. "Nonlinear gain control of dispersion-managed soliton amplitude and collisions". *Optical Fiber Technology.* Vol 6, No 2, 109-121. (2000).
9. P. M. Lushnikov. "Dispersion-managed solitons in optical fibers with zero average dispersion". *Optics Letters.* Vol 25, No 16, 1144-1146. (2000).
10. T. R. Wolinski. "Polarimetric optical fibers and sensors". *Progress in Optics.* Vol XL, 1-75. (2000).

Chapter 6
Optical Crosstalk

Modern all-optical networks that employ all-optical switches and optical add-drop multiplexers (OADMs) provide a tremendous high speed of several tens of Tera-bits per second. Such networks are challenging with many issues that have to be anticipated among which optical crosstalk is an important one.

Almost every component in an optical communication system may introduce one of the two different types of crosstalks which are known as in-band or out-of-band crosstalks. In optical networks, using wavelength division multiplexing (WDM), it is very common to use the terminologies of intra-channel and inter-channel respectively for in-band and out-of-band. The main difference between such crosstalk signals lies in their wavelength values. Although optical components are getting better by rejecting more than 35 dB from their adjacent channels, there still exists residual signals. The feature becomes more significant if channel powers are unequal.

The current WDM technology involves multiplexing several hundreds of wavelengths and transporting them in a single fiber, thus achieving Tb/s in a total capacity. If a WDM is used, in a fiber radio network, for example, then each base station (BS) can be assigned a single wavelength. A WDM requires selective optical components that can multiplex or demultiplex channels (or drop and add channels). These components can also cause an optical crosstalk by not fully removing unwanted channels.

When the crosstalk and input signals are on the same channel or wavelength, or have close-valued wavelengths, it is an in-band crosstalk. This kind of crosstalk accumulates very fast while passing through optical switches and have more serious effect than the same kind of crosstalk in the detection stage [7]. The in-band crosstalk, however, can be either coherent (phase-correlated) or incoherent (not phase-correlated) with the signal considered [4]. If the signal crosstalk mixing takes place within the laser coherent length, then in-band crosstalk is classified as a coherent crosstalk. Otherwise the incoherent crosstalk will appear. The coherent crosstalk impacts signal more severely than the incoherent crosstalk. The out-of-band or inter-channel crosstalk occurs when the crosstalk signal has wavelength that is different from that of useful signals.

The optical crosstalk even though originated from imperfections of small components mostly tends to generate large performance degradations and thus needs extra attentions. Experimental studies of in-band and out-of-band crosstalks show that the out-of-band crosstalk can be efficiently removed by narrow-band filters and has no influence on signals. On the other hand,the in-band crosstalk results in closing of signal eye-diagram and increases bit-error rate (BER).

6.1 In-band crosstalk

The in-band crosstalk arises from the signal leakage within optical cross-connects (OXCS) mainly due to component imperfections. It can also arise from reflections but this is not a serious source because these reflections can be controlled. The presence of crosstalk signals at the switch output is very important and needs improvements.

A crosstalk signal may occur as a leakage of channel 1 which travels with the signal in channel 2 and remains present at the output of the multiplexer. The demultiplexer should be able to ideally separate all wavelengths to different optical fibers but due to the component imperfection, a small portion may leak into the adjacent channel. Throughout the stage of multiplexing all wavelengths into a single fiber, the small portion of optical signals that leaked into the adjacent channel will leak back into the output of single fibers. Along with different nonlinear effects and effects from higher order dispersion, the influence of the crosstalk on optical pulse propagation becomes more severe.

There are two different approaches in studying the influence of crosstalks on the optical pulse propagation. The first one is the analytical method that uses the linear model of a single mode fiber and the second one uses nonlinear effects and direct NLSE solving. The analytical method analyzes the joint influence of the higher order dispersion and crosstalk on optical pulse dynamics in a closed form.

The linear model of a single mode fiber with length L is given by its transfer function as [1]

$$H(\omega) = \exp\{-(\alpha + i\beta(\omega))L\} \tag{6.1}$$

where α is the attenuation coefficient, $\beta(\omega)$ is the phase constant that can be expressed in Taylor series about $\omega = \omega_0$ as

$$\beta(\omega) = \sum_{n=0}^{\infty} \frac{1}{n!} \left[\frac{d^n\beta(\omega)}{d\omega^n}\right]_{\omega=\omega_0} (\omega - \omega_0)^n \tag{6.2}$$

The impulsive response is obtained by using the inverse Fourier transform of $H(\omega)$:

$$\begin{aligned} h(t) &= \frac{1}{2\pi}\int_{-\infty}^{\infty} H(\omega)e^{i\omega t}d\omega \\ &= \frac{1}{2\pi}e^{-\alpha L}e^{i(\omega_0 t-\beta_0 L)}\int_{-\infty}^{\infty}\exp\left[i\left\{\omega\left(t-\beta_1 L\right)\right.\right. \\ &\quad \left.\left.-\frac{1}{2}\beta_2\omega^2 L-\frac{1}{6}\beta_3\omega^3 L-\cdots\right\}\right]d\omega \end{aligned} \tag{6.3}$$

Applying Euler's lemma to Eq.(6.3) simplifies it to

$$h(t) = \frac{1}{2\pi}e^{-\alpha L}\exp\left[i\left\{\omega_0 t-\beta_0 L+\theta(t)\right\}\right]\sqrt{I_c^2(t)+I_s^2(t)} \tag{6.4}$$

where

$$I_c(t) = \int_{-\infty}^{\infty}\cos\left(\omega t-\sum_{n=1}^{\infty}\frac{1}{n!}\beta_n L\omega^n\right)d\omega \tag{6.5}$$

$$I_s(t) = \int_{-\infty}^{\infty}\sin\left(\omega t-\sum_{n=1}^{\infty}\frac{1}{n!}\beta_n L\omega^n\right)d\omega \tag{6.6}$$

$$\theta(t) = \tan^{-1}\frac{I_s(t)}{I_c(t)} \tag{6.7}$$

The characteristics of phase shifts, denoted by $\theta(t)$, are induced by the higher order dispersion terms. the impulsive response $h(t)$ corresponds to the electrical field in physical sense and its envelope can be expressed as

$$|h(t)|^2 = \frac{1}{4\pi^2}\left[I_c^2(t)+I_s^2(t)\right] \tag{6.8}$$

General forms of optical signals in medium with a dominant dispersion effect is obtained by using Eqs.(6.4) and (6.7) with the integrals $I_c(t)$ and $I_s(t)$ defined as

$$I_1(t) = \int_{-\infty}^{\infty}F(\omega)\cos\left(\omega t-\sum_{n=1}^{\infty}\frac{1}{n!}\beta_n L\omega^n\right)d\omega \tag{6.9}$$

$$I_2(t) = \int_{-\infty}^{\infty}F(\omega)\sin\left(\omega t-\sum_{n=1}^{\infty}\frac{1}{n!}\beta_n L\omega^n\right)d\omega \tag{6.10}$$

where $F(\omega)$ represents the Fourier transform of input signals.

6.2 Gaussian optical pulse

Pulses from many lasers can be approximated by Gaussian pulses and the most important feature of such pulses is that they maintain their shape during propagation. Therefore, the input optical signal with a Gaussian envelope is represented by

$$s_1(0,t) = \sqrt{P_0} e^{-(t/T_0')^2 + i\omega_0 t} \tag{6.11}$$

where P_0 is the optical pulse peak power, $T_0' = T_0/\sqrt{2}$ is pulse half-width (at $1/e$ intensity point) and ω_0 is the optical carrier frequency.

The in-band interference is at the same frequency as an useful signal but the time and phase shifted in regard to the useful signal, as

$$s_2(0,t) = \sqrt{P_i} e^{-(t/T_0' - b')^2 + i(\omega_0 t + \phi)} \tag{6.12}$$

where P_i is the interference peak power, b ($b = b'T_0$) and ϕ are the time and phase shift, respectively. The value of the timing shift, or the propagation delay, depends on the nature of the in-band interference (coherent crosstalk or multi-path reflection). The phase shift ϕ varies in a random manner due to temperature and wavelength variations in the range $[0, \pi]$. The envelope and phase of the resulting signal $s_r(t)$ at the fiber input are [8]

$$|s_r(t)| = \Big[P_0 e^{-2(t/T_0')^2} + 2\sqrt{P_0 P_i} e^{-(t/T_0')^2} e^{-(t/T_0' - b')^2} \cos\phi \\ + P_i e^{-2(t/T_0' - b')^2} \Big]^{1/2} \tag{6.13}$$

$$\psi(t) = \tan^{-1} \frac{\sqrt{P_i} e^{-(t/T_0' - b')^2} \sin\phi}{\sqrt{P_0} e^{-(t/T_0')^2} + \sqrt{P_i} e^{-(t/T_0' - b')^2} \cos\phi} \tag{6.14}$$

The in-band interference can be constructive or destructive, depending on the phase-shift value. If the worst case is considered, i.e., the destructive interference and if it is assumed that it appears at the beginning of an optical fiber (e.g. a double reflection [3] or an in-band crosstalk resulting from WDM components used for routing and switching along optical networks [5])the receiver pulse shape under the influence of n-th order dispersion is [9]

$$r_n(t,L) = \frac{\sqrt{P_0} T_n}{\sqrt{\pi}} e^{-\alpha L} e^{i\{\omega_0 t - \beta_0 L + \theta(\tau)\}} \sqrt{I_1^2(\tau) + I_2^2(\tau)} \tag{6.15}$$

$$\theta(\tau) = \tan^{-1} \frac{I_2(\tau)}{I_1(\tau)} \tag{6.16}$$

with

$$I_1(\tau) = \int_{-\infty}^{\infty} e^{-u^2 T_n^2} \left[\left\{ 1 - \sqrt{\frac{P_i}{P_0}} \cos(2b' T_n u) \right\} \cos(u\tau - u^n) \right.$$

$$-\sqrt{\frac{P_i}{P_0}}\sin(2b'T_n u)\sin\left(u\tau - u^n\right)\Bigg] du \tag{6.17}$$

$$I_2(\tau) = \int_{-\infty}^{\infty} e^{-u^2T_n^2}\left[\left\{1-\sqrt{\frac{P_i}{P_0}}\cos(2b'T_n u)\right\}\sin\left(u\tau - u^n\right)\right.$$

$$\left.-\sqrt{\frac{P_i}{P_0}}\sin(2b'T_n u)\cos\left(u\tau - u^n\right)\right] du \tag{6.18}$$

where Eqs.(6.17) and (6.18) are obtained by changing the following parameters in Eqs.(6.9) and (6.10):

$$b_n = \left(\frac{\beta_n L}{n!}\right)^{-1/n}, \quad \omega = b_n u, \quad \tau = b_n t, \quad T_n = \frac{b_n T_0}{2} \tag{6.19}$$

It it is assumed that binary data sequences with values 0 and 1 are transmitted along the optical fiber. For non-overlapping of light pulses down an optical fiber link, the digital bit rate (B) must be less than the reciprocal of the broadened (through certain order of dispersion) pulse duration. Depending on the interference time shift value, the coherent interference can be left ($b < 0$) or right ($b > 0$) shifted from the center of data in binary sequences. The propagation length can be expressed via dispersion length that is for n-th order dispersion is given by

$$L_D = \frac{(T_0')^n}{|\beta_n|} \tag{6.20}$$

The second order dispersion induces a symmetrical broadening. The greater time-shift of interference induces the asymmetrical pulse deformation. Note that both, the noisy nature of the input to a clock-recovery circuit and noise produced by optical amplifiers, timing jitter can be induced. Then, as previously mentioned,the asymmetrical pulse deformation can be dangerous. The worst case in the detection process is $b = 0$. This situation is very often seen in switching systems [1, 2]. The eye diagram for the worst case is shown in Figure 6.1(a).

The crosstalk level is defined by the signal-to-noise ratio (SIR) i.e., the ratio of useful signal optical power to crosstalk signal optical power. It is defined as

$$\text{SIR} = 20\log\frac{P_0}{P_i} \tag{6.21}$$

If one of the methods that compensate degrading influences of the second order dispersion is employed, the third order dispersion remains and has a greater influence on the pulse shape. The third order dispersion distorts the pulse shape such that it becomes asymmetrical with an oscillatory structure near one of its edges ($\beta_3 > 0$ affects the trailing edge of the pulse while

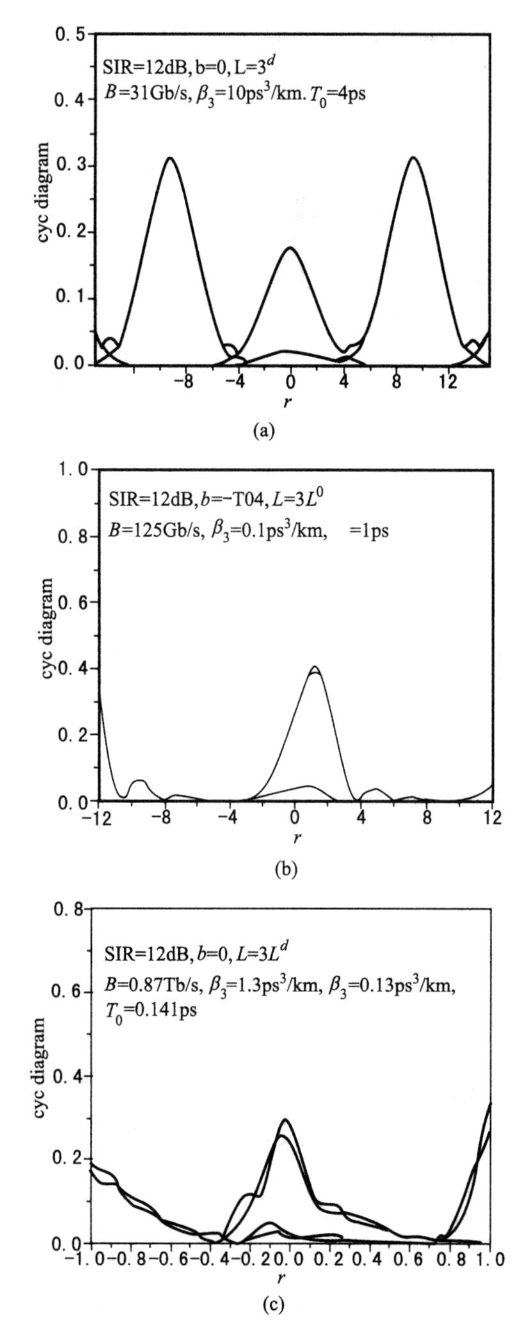

Fig. 6.1 Eye diagram for (a) second order dispersion, (b) third order dispersion, (c) second and third order dispersion in the presence of coherent interference.

$\beta_3 < 0$ affects the leading edge of the pulse). Because of the asymmetrical deformation of pulses induced by the third order dispersion (oscillation on the trailing edge) the biggest error in the detection process will occur for small negative interference time shifts. The effect of a negative interference is more destructive than a positive interference at the receiving end of the fiber due to the timing shift of the resulting pulse. The opposite situation will happen for $\beta_3 < 0$. Great absolute values of time shifts can increase an inter-symbol interference (ISI) if the transmission rate is high enough to induce a sizeable overlapping of pulses. Figure 6.1(b) shows the eye diagram for the case of the third order dispersion, when the second order dispersion is suppressed in the presence of the most destructive interference.

There is a case (equal dispersion length for the second and the third order dispersions) when it is needed to investigate the joint influence of the second and the third order dispersions on pulse propagation along the linear optical fiber. Then analytical expressions describing the pulse shape along the optical fiber have the following form [8]:

$$r_{2,3}(t,L) = \frac{\sqrt{P_0}T_0}{2\sqrt{\pi}} e^{-\alpha L} e^{i\{\omega_0 t - \beta_0 L + \theta(\tau)\}} \sqrt{I_1^2(t) + I_2^2(t)} \tag{6.22}$$

$$\theta(t) = \tan^{-1}\frac{I_2(t)}{I_1(t)} \tag{6.23}$$

where

$$I_1(t) = \int_{-\infty}^{\infty} e^{-\frac{\omega^2 T_0^2}{4}} \left[\left\{ 1 - \sqrt{\frac{P_i}{P_0}} \cos\left(b' T_0 \omega\right) \right\} \cos(\omega t - b_2\omega^2 - b_3\omega^3) - \sqrt{\frac{P_i}{P_0}} \sin\left(b' T_0 \omega\right) \sin(\omega t - b_2\omega^2 - b_3\omega^3) \right] d\omega \tag{6.24}$$

$$I_2(t) = \int_{-\infty}^{\infty} e^{-\frac{\omega^2 T_0^2}{4}} \left[\left\{ 1 - \sqrt{\frac{P_i}{P_0}} \cos\left(b' T_0 \omega\right) \right\} \sin(\omega t - b_2\omega^2 - b_3\omega^3) + \sqrt{\frac{P_i}{P_0}} \sin\left(b' T_0 \omega\right) \cos(\omega t - b_2\omega^2 - b_3\omega^3) \right] d\omega \tag{6.25}$$

and

$$b_2 = \left(\frac{\beta_2 L}{2!}\right), \quad b_3 = \left(\frac{\beta_3 L}{3!}\right) \tag{6.26}$$

Figure 6.1(c) shows the joint influence of the second and the third order dispersions and in-band crosstalks.The Gaussian pulse at the receiver is broadened by the second order dispersion and it has a long trailing edge as a result of the third order dispersion influence. Because of such pulse deformation, the position of in-band crosstalk signals in regard to center of a bit is very important. In a detection process,the bigger error is made in the following cases.

1. If $|b| < T_0$, for in-band crosstalk signal right shifted with respect to the center of bit.
2. If $|b| \geq T_0$, for in-band crosstalk with respect to the center of bit.

The results presented in Figure 6.2 below testify that the pulse becomes most distorted for the case when there is no time shift for in-band crosstalk signals. The eye diagram for this case is shown in Figure 6.1(c).

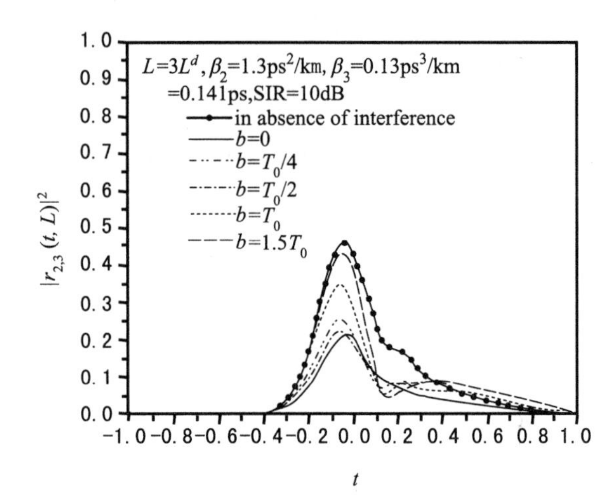

Fig. 6.2 The pulse shape at the end of the optical fiber ($L = 3L^d$) under the joint influence of the second and the third order dispersions for SIR = 10 dB.

6.2.1 Bit error rate

When the transmission rate (B) and the transmission distance (L) is fixed, a suitable measure of the line performance is the bit error rate (BER). Since the BER has to be extremely small, with numerical tools, it is very difficult and time consuming to perform full simulations of the system in order to determine BER simply by counting the mistakes. Therefore, it is of great interest to find a proper statistical approximation of the BER.

The most commonly used technique to evaluate the Intensity Modulation-Direct Detection (IM-DD) system performance assumes a Gaussian white noise distribution on both the zero and the one levels [10]. It is difficult to detect interference time shifts especially when there are many connections and taps in the system because they may cause reflections too. Therefore, it is treated as a random variable with a uniform probability density function

$p(b)$. The BER, in this case, is given by

$$\text{BER} = \frac{1}{2}\int_{-\frac{1}{2B}}^{\frac{1}{2B}} \left[\int_{i_D}^{+\infty} \frac{1}{\sqrt{2\pi}(\overline{i_N^2})^{1/2}} \exp\left\{ -\frac{(i - i_{\text{sig}_0})^2}{2\overline{i_N^2}} \right\} di \right.$$
$$\left. + \int_{-\infty}^{i_D} \frac{1}{\sqrt{2\pi}(\overline{i_N^2})^{1/2}} \exp\left\{ -\frac{(i - i_{\text{sig}_1})^2}{2\overline{i_N^2}} \right\} di \right] p(b)db \tag{6.27}$$

where i_{sig_0} and i_{sig_1} are the mean value of the currents, in 0 and 1 states respectively, i_D is the decision threshold and $\overline{i_N^2}$ is the mean square noise current. The BER for joint influence of the second and the third order dispersions in the presence of coherent interference is shown in Figure 6.3.

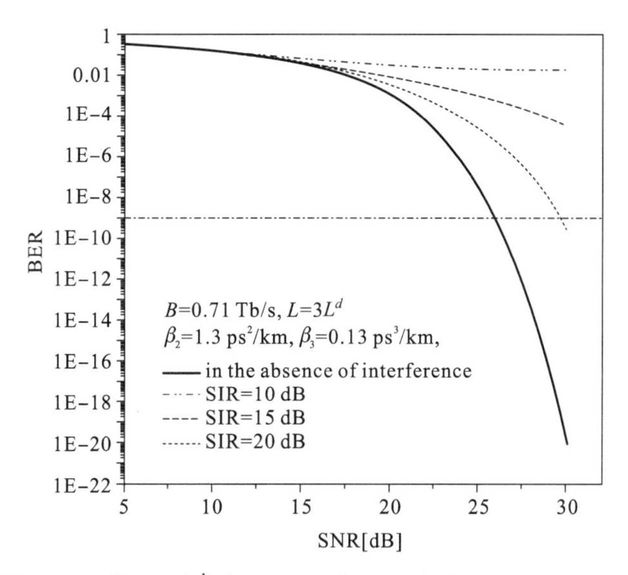

Fig. 6.3 BER curves ($L = 3L^d$) for joint influence of the second and the third order dispersions.

6.3 Sech optical pulse

The sech model of the input optical pulse and the in-band interference (crosstalk) will now be considered that has following shapes:

$$s(t) = \frac{\sqrt{P_0}}{\cosh(t/T_0)} e^{i\omega_0 t} \tag{6.28}$$

$$s_i(t) = \frac{\sqrt{P_i}}{\cosh(\frac{t}{T_0} - b')} e^{i(\omega_0 t + \phi)} \tag{6.29}$$

where P_i is the interference peak power and b ($b = b'T_0$) is the interference time shift. The phase shift ϕ varies in a random manner due to the temperature and wavelength variations in the range $(0, \pi)$. The envelope and phase of the resulting signal $s_r(t)$ at the fiber input are [6]

$$|s_r(t)| = \left[\frac{P_0}{\cosh^2 \frac{t}{T_0}} + \frac{2\sqrt{P_0 P_1} \cos\phi}{\cosh \frac{t}{T_0} \cosh(\frac{t}{T_0} - b')} + \frac{P_i}{\cosh^2(\frac{t}{T_0} - b')} \right]^{1/2} \tag{6.30}$$

$$\psi(t) = \tan^{-1} \frac{\frac{\sqrt{P_i} \sin\phi}{\cosh(\frac{t}{T_0} - b')}}{\frac{\sqrt{P_0}}{\cosh \frac{t}{T_0}} + \frac{\sqrt{P_i} \cos\phi}{\cosh(\frac{t}{T_0} - b')}} \tag{6.31}$$

A general expression for the fiber response for an arbitrary input pulse is already given in Eq.(6.15), which for the case with the influence of the second and the fourth order dispersions, can be written as in Eqs.(6.22) and (6.23) where now

$$\begin{aligned} I_1(t) = \int_{-\infty}^{\infty} \frac{1}{\cosh \frac{\pi T_0 \omega}{2}} & \left[\left\{ 1 - \sqrt{\frac{P_i}{P_0}} \cos(b' T_0 \omega) \right\} \right. \\ & \cdot \cos(\omega t - b_2 \omega^2 - b_3 \omega^3) \\ & \left. - \sqrt{\frac{P_i}{P_0}} \sin(b' T_0 \omega) \sin(\omega t - b_2 \omega^2 - b_3 \omega^3) \right] d\omega \end{aligned} \tag{6.32}$$

$$\begin{aligned} I_2(t) = \int_{-\infty}^{\infty} \frac{1}{\cosh \frac{\pi T_0 \omega}{2}} & \left[\left\{ 1 - \sqrt{\frac{P_i}{P_0}} \cos(b' T_0 \omega) \right\} \right. \\ & \cdot \sin(\omega t - b_2 \omega^2 - b_3 \omega^3) \\ & \left. + \sqrt{\frac{P_i}{P_0}} \sin(b' T_0 \omega) \cos(\omega t - b_2 \omega^2 - b_3 \omega^3) \right] d\omega \end{aligned} \tag{6.33}$$

Binary data sequences with values 0 and 1 are transmitted through the optical fiber. The digital bit rate (B) is less than the reciprocal of the broadened pulse duration. Depending on the value of interference time shifts, the coherent interference can be left ($b < 0$) or right ($b > 0$) shifted in regard to center of data in binary sequences. The propagation length is expressed via the dispersion length which for the nth order dispersion is given by Eq.(6.20).

The pulse shape under the influence of the second and the fourth order dispersions is shown in Figure 6.4. Strong influence of the interference may be noticed even for $b > T_0$ and long trailing ends will unquestionably induce

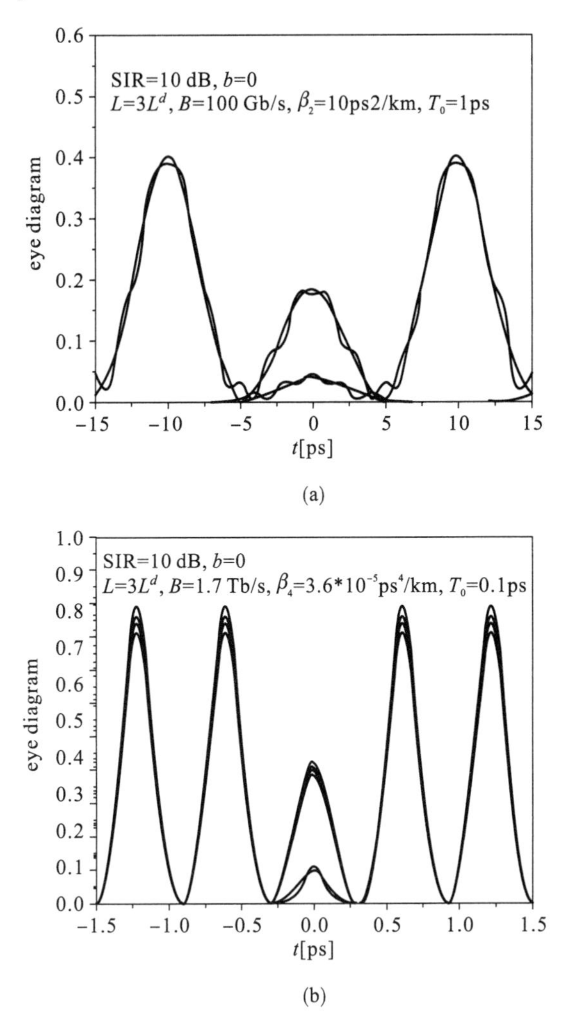

Fig. 6.4 Eye diagram for individual influence of (a)the second order dispersion and (b) the fourth order dispersion in the presence of worst case interferences.

the inter-symbol interference (ISI). The worst case in the detection process happens for $b = 0$ and is seen in Figure 6.5 below.

The joint influence of the second and the fourth order dispersions is shown in Figure 6.6.

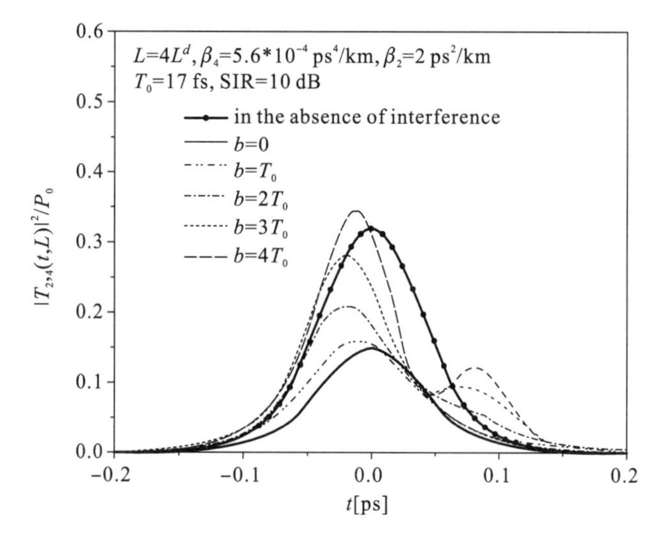

Fig. 6.5 Pulse shape at the end of the optical fiber ($L = 4LD$) under the second and the fourth order dispersions for SIR = 10 dB and different time shifts $b > 0$.

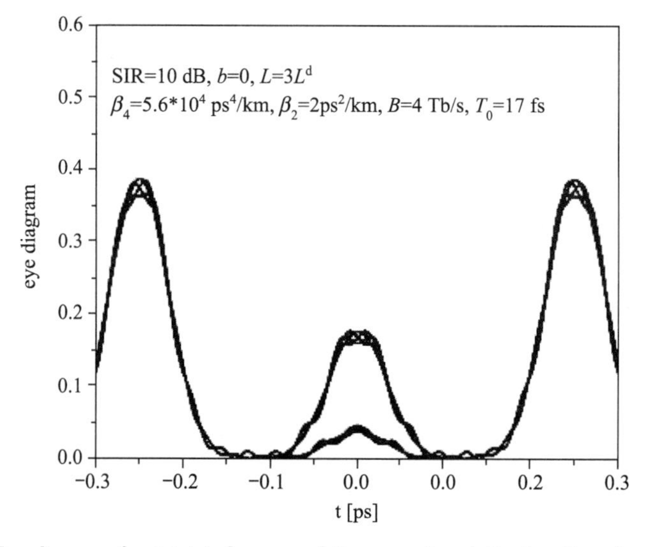

Fig. 6.6 Eye diagram for joint influences of the second and the fourth order dispersions ($L = 3L^d$) in the presence of the worst case interference.

6.4 Super-Sech optical pulse

A more general form of the sech pulse (and interference) can be mathematically represented as a super-sech pulse model having following forms:

$$s(t) = \frac{\sqrt{P_0}}{\cosh^m \frac{t}{T_0}} e^{i\omega_0 t} \tag{6.34}$$

$$s_i(t) = \frac{\sqrt{P_i}}{\cosh^m (\frac{t}{T_0} - b')} e^{i(\omega_0 t + \phi)} \tag{6.35}$$

The Fourier transform $F(\omega)$ of the input pulse for even m ($m = 2k$ where $k = 0, 1, 2, \ldots$) can be written as

$$F(\omega) = 2^m T_0 \left[\frac{F(m, \frac{m - iT_0\omega}{2}, 1 + \frac{m - iT_0\omega}{2}, -1)}{m - iT_0\omega} + \frac{F(m, \frac{m + iT_0\omega}{2}, 1 + \frac{m + iT_0\omega}{2}, -1)}{m + iT_0\omega} \right] \tag{6.36}$$

where the Gauss' hyper-geometric function is defined as

$$F(\alpha, \beta; \gamma; z) = \frac{\Gamma(\gamma)}{\Gamma(\alpha)\Gamma(\beta)} \sum_{n=0}^{\infty} \frac{\Gamma(\alpha + n)\Gamma(\beta + n)}{\Gamma(\gamma + n)} \frac{z^n}{n!} \tag{6.37}$$

while if m is odd ($m = 2k + 1$ where $k = 0, 1, 2, \ldots$), the Fourier transform is given by

$$F(\omega) = \frac{1}{\sqrt{\pi}} 2^{\frac{m}{2} - 2} e^{-\frac{\pi}{4}\{(k-1)i + 2\omega\}} \left[B_{-1}\left(\frac{1}{4}(k - 1 - 2i\omega), \frac{1}{2}(3 - k)\right) + e^{\pi\omega} B_{-1}\left(\frac{1}{4}(k - 1 + 2i\omega), \frac{1}{2}(3 - k)\right)\right] \tag{6.38}$$

where the incomplete beta function is defined as

$$B_x(a, b) = \int_0^x t^{a-1}(1 - t)^{b-1} dt \tag{6.39}$$

Now, both the Gauss' hyper-geometric function and the incomplete beta function can be numerically evaluated with any arbitrary precision. In Figures 6.7(a) and (b), the super-sech optical pulse and crosstalk with $m = 2$ are shown.

A more realistic scenario includes nonlinearities in optical fibers along with the dispersion. In such case, the optical fiber transfer function (6.1) is no longer valid. The influence of the crosstalk on an optical pulse propagation can be determined by solving the NLSE for the in-band crosstalk or by solving a set of coupled NLSE for the out-of-band crosstalk.

The in-band crosstalk and useful optical signal are approximately at the same frequencies:

$$s_1(0, t) = A_1(0, t) \cos(\omega_0 t) \tag{6.40}$$

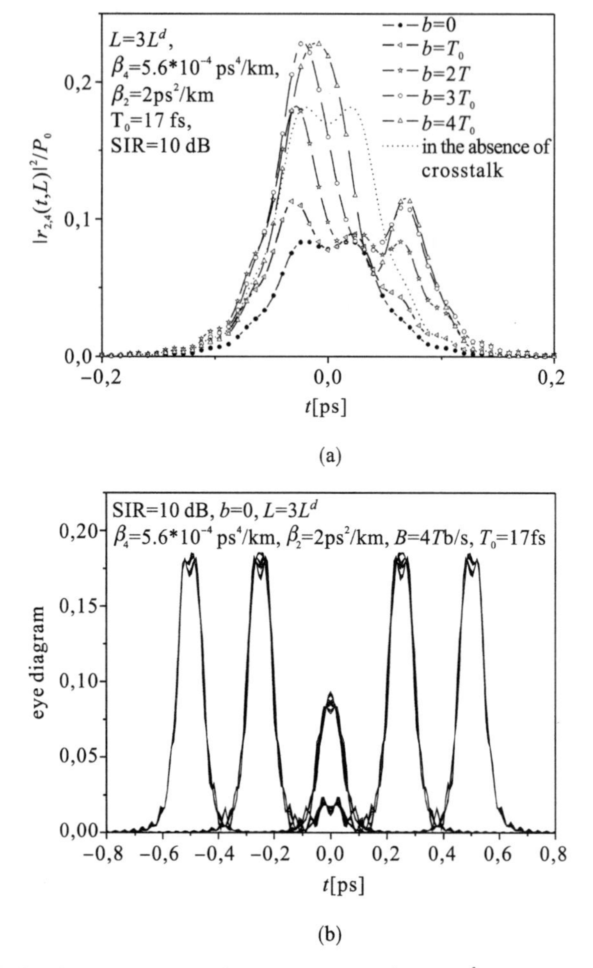

Fig. 6.7 (a) Pulse shape at the end of the optical fiber ($L = 3L^d$) under the second and the fourth order dispersions for SIR=10 dB and different time shifts $b > 0$, (b) corresponding eye-diagram.

$$s_2(0,t) = A_2(0,t)\cos(\omega_0 t) \tag{6.41}$$

where the pulse propagation is governed by the NLSE that is given by

$$iA_z = -\frac{i\alpha}{2}A - \frac{\beta_2}{2}A_{tt} - \gamma|A|^2 A \tag{6.42}$$

with

$$\begin{aligned} A &= A_r(0,T) \\ &= \sqrt{A_1^2(0,T) + 2A_1(0,T)A_2(0,T)\cos\phi + A_2^2(0,T)} \end{aligned} \tag{6.43}$$

In this case, the crosstalk and useful optical signal will propagate superimposed through an optical fiber.

The propagation of short pulses under the influence of the out-of-band crosstalk being at a different wavelength than useful signals is governed by a set of two coupled NLSE:

$$A_{1_z} + \frac{i\beta_{21}}{2} A_{1_{tt}} = i\gamma_1(|A_1|^2 + 2|A_2|^2)A_1 \tag{6.44}$$

$$A_{2_z} + \frac{i\beta_{22}}{2} A_{2_{tt}} = i\gamma_1(|A_2|^2 + 2|A_1|^2)A_2 \tag{6.45}$$

where

$$\gamma_j = \frac{n_2\omega_j}{cA_{\text{eff}}}$$

$$\beta_{2j} = -\frac{D\Lambda_j^2}{2\pi c}$$

where $j = 1, 2$ and

$$T = t - \frac{z}{v_g}$$

$$\gamma = \frac{n_2\omega_0}{cA_{\text{eff}}}$$

Here γ is the coefficient of nonlinearity and $A_{\text{eff}} = \pi w^2$ is the effective core area. A_{eff} is typically 10^{-20} μm^2 in the visible region but can be in the range 50–80 μm^2 in the 1.55 μm region, so γ can vary over the range 2–30 $\text{W}^{-1}\cdot\text{km}^{-1}$ depending on n_2.

Each of these two coupled NLSE in Eqs.(6.44) and (6.45) is a nonlinear partial differential equation which can be solved by using the split-step Fourier method (SSFM) [1]. The input or useful optical pulse and the out-of-band crosstalk signal can be modeled, respectively, as

$$s_1(0,t) = A_1(0,t)\cos(\omega_1 t) \tag{6.46}$$

$$s_2(0,t) = A_2(0,t)\cos(\omega_2 t) \tag{6.47}$$

where

$$A_1(0,t) = \sqrt{P_1} f\left(-\frac{t^2}{2T_0^2}\right) \tag{6.48}$$

$$A_2(0,t) = \sqrt{P_2} f\left(-\frac{(t-T_s)^2}{2T_0^2}\right) \tag{6.49}$$

where P_1 and P_2 are the peak powers of the useful optical pulse and the out-of-band crosstalk, respectively and T_s is the out-of-band crosstalk time shift. Also, f represents the pulse shape. It could be Gaussian, super-Gaussian or super-sech as the case may be. It is assumed that $|T_s| < T_b/2$ since the interest is confined to one bit period T_b.

The pulse evolution picture contour plot for the case of the super-Gaussian pulse is shown in Figure 6.8. The following factors are taken into consideration: $T_{\mathrm{FWHM}} = 12.5\,\mathrm{ps}$, $\lambda_1 = 1550\,\mathrm{nm}$, bit rate $R = 20\,\mathrm{Gb/s}$, $P_1 = 50\,\mathrm{mW}$, SMF is the regime of a normal dispersion ($D = 0.2\,\mathrm{ps/nm\text{-}km}$) with a parameter $A_{\mathrm{eff}} = 50\,\mu\mathrm{m}^2$. The out-of-band crosstalk wavelength is $\lambda_2 = 1551.5\,\mathrm{nm}$. The fiber length is 60 km and SIR = 0 dB.

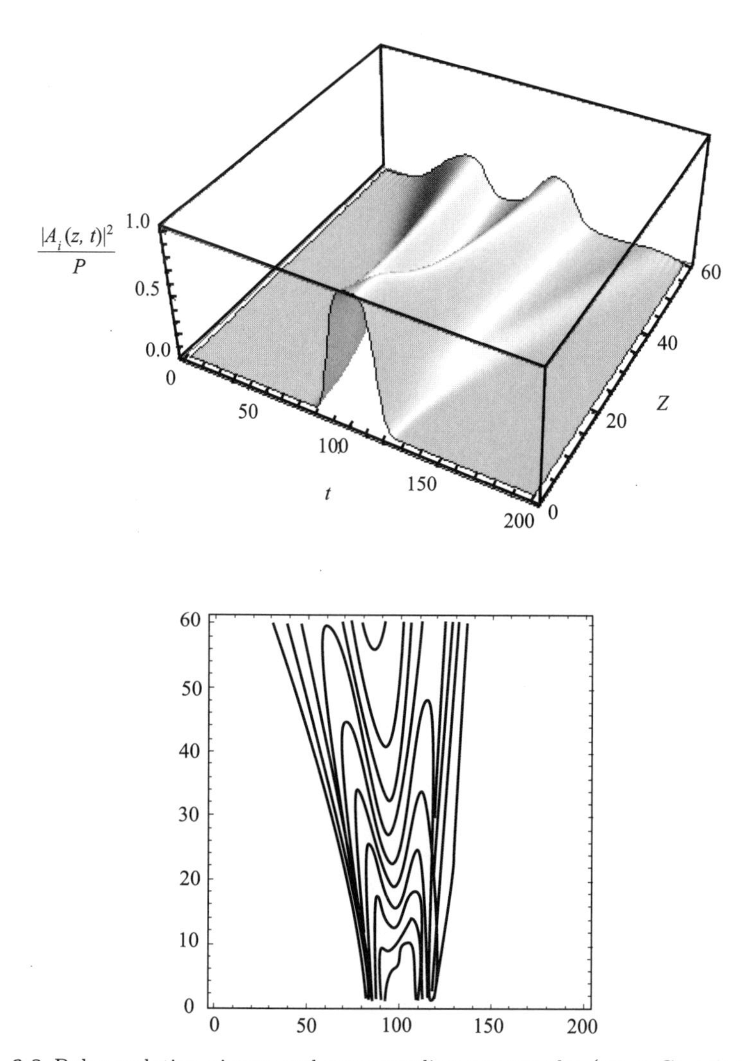

Fig. 6.8 Pulse evolution picture and corresponding contour plot (super-Gaussian pulse).

The influence of the out-of-band crosstalk occurring at the input of the transmission link can be expressed by estimating the eye opening penalty (EOP). The EOP is a performance measure used very often when considering the dynamic propagation effects, such as the dispersion, nonlinearities and other influences that distort the pulse shape.The EOP is defined as a ratio of the initial eye opening (EO_{before}) to the eye opening after transmission (EO_{after}). The initial eye opening is measured at the fiber input:

$$\mathrm{EOP} = 20 \log \frac{\mathrm{EO_{before}}}{\mathrm{EO_{after}}} \tag{6.50}$$

The out-of-band crosstalk occurring at the fiber input greatly affects the eye opening. The change of the EOP with a crosstalk time-shift and the SIR is shown in Figure 6.9. For small values of time-shifts, The EOP has large values which means that the crosstalk induces greater eye closing. When the out-of-band crosstalk optical power is equal to the signal optical power (SIR = 0), the nonlinear effects additionally distort the pulse shape and EOP becomes greater. A thorough analysis of the crosstalk influence is very useful for improving the existence transmission links or designing the new ones. It is also very useful in designing the DWDM systems as this type of crosstalks is pretty common.

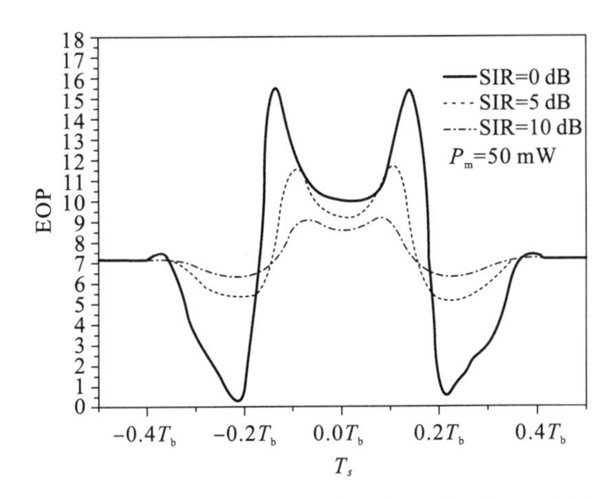

Fig. 6.9 EOP vs Ts for super-Gaussian optical pulse and different SIR.

References

1. G. P. Agrawal. *Nonlinear Fiber Optics*. Academic Press, San Deigo, CA. USA. (1995).
2. A. Ehrhardt, M. Eiselt, G. Goßkopf, L. Küller, W. Pieper, R. Schnabel, G. H. Weber. "Semiconductor laser amplifier as optical switching gate". *Journal of Lightwave Technology*. Vol 11, No 8, 1287-1295. (1993).
3. A. D. Fishman, G. D. Duff & A. J. Nagel. "Measurement and simulation of multipath interference for 1.7 Gb/s lightwave transmission systems using single- and multi-

frequency laser". *Journal of Lightwave Technology.* Vol 8, No 6, 894-905. (1990).

4. A-H. Guan & Y-H. Wang. "Experimental study of interband and intraband crosstalk in WDM networks". *Optoelectronics Letters.* Vol 4, No 1, 42-44. (2008).
5. E. Iannone, R. Sabella, M. Avattaneo & G. De Paolis. "Modeling of in-band crosstalk in WDM optical networks". *Journal of Lightwave Technology.* Vol 17, No 7, 1135-1141. (1999).
6. A. Panajotovic, D. Milovic & A. Mittic. "Boundary case of pulse propagation analytic solution in the presence of interference and higher order dispersion". *TELSIKS 2005 Conference Proceedings.* 547-550. Nis-Serbia. (2005).
7. Y. Pointurier, M. Brandt-Pearce & C. L. Brown. "Analytical study of crosstalk propagation in all-optical networks using perturbation theory". *Journal of Lightwave Technology.* Vol 23, No 12, 4074-4083. (2005).
8. M. Stefanovic, D. Draca, A. Panajotovic & D. Milovic. "Individual and joint influence of second and third order dispersion on transmission quality in the presence of coherent interference". *Optik.* Vol 120, No 13, 636-641.(2009).
9. B. Stojanovic, D. M. Milovic & A. Biswas. "Timing shift of optical pulses due to inter-channel crosstalk". *Progress in Electromagnetics Research M.* Vol 1, 21-30. (2008).
10. J. M. Senior. *Optic Fiber Communications.* Prentice Hall, New York, NY. USA. (1992).

Chapter 7
Gabitov-Turitsyn Equation

7.1 Introduction

This chapter is devoted to the study of the DM-NLSE in polarization preserving fibers, birefringent fibers as well as DWDM systems by the aid of multiple scale analysis. When this technique applied to the DM-NLSE it will convert the nonlinear partial differential equation to a nonlinear integro-differential equation with a nonlinear non-local kernel. This integro-differential equation is known as the Gabitov-Turitsyn equation (GTE) that first appeared in 1996 [16]. Later in 1998, this equation was refined in a simpler format by Ablowitz and Biondini [3]. Later, this equation was extended by Biswas to the cases of birefringent fibers and DWDM systems in 2001 and 2003, respectively [11–13].

The GTE will be the universal asymptotic equation that governs the evolution of the amplitude of an optical pulse for a dispersion-managed system that is governed by Eq.(2.41). In the GTE all fast and large variations are removed. It is necessary to note that GTE is equally applicable to the case of pulse dynamics for a zero or normal value of the average dispersion. The special solution will be considered with Gaussian pulses in a separate subsection. The case of super-Gaussian and super-sech pulses are yet to be studied at this point. Finally, the result will be extended to the case of birefringent fibers and DWDM systems. The starting point is same equation as in Eq.(2.41) which is

$$iu_z + \frac{D(z)}{2}u_{tt} + g(z)|u|^2u = 0 \tag{7.1}$$

This equation will be studied asymptotically now in the following section.

7.2 Polarization-preserving fibers

This section has been taken from the work of Ablowitz and Biondini that appeared in 1998 [3]. Equation (7.1) contains both large and rapidly varying terms. To obtain the asymptotic behavior, the fast and slow z scales are introduced respectively as

$$\zeta = \frac{z}{z_a} \qquad \text{and} \qquad Z = z$$

Now, expand the field u in powers of z_a [6, 30, 31]:

$$u(\zeta, Z, t) = u^{(0)}(\zeta, Z, t) + z_a u^{(1)}(\zeta, Z, t) + z_a^2 u^{(2)}(\zeta, Z, t) + \cdots \tag{7.2}$$

Again, decompose Eq. (7.1) into a series of equations corresponding to the different powers of z_a. In general, at $O(z_a^{n-1})$ yields

$$F\left[u^{(n)}\right] = -P_n\left[u^{(0)}, u^{(1)}, \ldots, u^{(n-1)}\right] \tag{7.3}$$

where

$$F\left[u^{(n)}\right] = iu_\zeta^{(n)} + \frac{1}{2}\Delta(\zeta)u_{tt}^{(n)} \tag{7.4}$$

with

$$P_0 = 0 \tag{7.5}$$

$$P_1\left[u^{(0)}\right] = iu_Z^{(0)} + \frac{1}{2}\delta_a u_{tt}^{(0)} + g(z)\left|u^{(0)}\right|^2 u^{(0)} \tag{7.6}$$

while

$$P_2\left[u^{(0)}, u^{(1)}\right] = iu_Z^{(1)} + \frac{1}{2}\delta_a u_{tt}^{(1)} + g(z)\left[2|u^{(0)}|^2 u^{(1)} + (u^{(0)})^2 (u^{(1)})^*\right] \tag{7.7}$$

and so on. The operators L and F are defined as

$$L[f] \equiv i\frac{\partial f}{\partial \zeta} + \frac{\Delta(\zeta)}{2}\frac{\partial^2 f}{\partial t^2}$$

$$F[f, h] \equiv i\frac{\partial f}{\partial z} + \frac{\delta_a}{2}\frac{\partial^2 f}{\partial t^2} + g(z)|f|^2 f$$

So, $O(1/z_a)$ gives

$$L[u^{(0)}] = 0 \tag{7.8}$$

while $O(1)$ gives

$$L[u^{(1)}] + F[u^{(0)}] = 0 \tag{7.9}$$

The Fourier transform and its inverse are respectively defined as

$$\hat{f}(\omega) = \mathbf{F}[f] \equiv \int_{-\infty}^{\infty} f(t)e^{i\omega t}dt \tag{7.10}$$

$$f(t) = \mathbf{F}^{-1}[\hat{f}] \equiv \frac{1}{2\pi}\int_{-\infty}^{\infty} \hat{f}(\omega)e^{-i\omega t}d\omega \tag{7.11}$$

In this chapter, the convolution theorem of the Fourier transform will be used. The inner product $f * g$ of the two functions $f(x)$ and $g(x)$ is defined as the integral

$$(f * g)(x) = \int_{-\infty}^{\infty} f(x')g(x-x')dx' \tag{7.12}$$

and the convolution theorem states that

$$\mathbf{F}^{-1}[\hat{f}\hat{g}] = f * g \tag{7.13}$$

and

$$\mathbf{F}[fg] = \frac{1}{2\pi}\hat{f} * \hat{g} \tag{7.14}$$

Taking the Fourier transform of Eq.(7.8) gives

$$-i\hat{L}[u^{(0)}] \equiv \frac{\partial \hat{u}^{(0)}}{\partial \zeta} + \frac{i\omega^2}{2}\Delta(\zeta)\hat{u}^{(0)} = 0 \tag{7.15}$$

whose solution is

$$\hat{u}^{(0)}(\zeta, Z, \omega) = \hat{U}(Z,\omega)\,\hat{p}\,(C(\zeta),\omega) \tag{7.16}$$

where

$$\hat{p}(x,\omega) = e^{-\frac{i}{2}x\omega^2} \qquad \text{and} \qquad C(\zeta) = C_0 + \int_0^{\zeta} \Delta(\zeta')d\zeta'$$

with arbitrary C_0. The integration constant $\hat{U}(Z,\omega)$ represents the slowly varying amplitude of $u^{(0)}$, while $\hat{p}\,(C(\zeta),\omega)$ contains the fast periodic oscillations due to large local values of the dispersion. The function $\hat{U}(Z,\omega)$ is arbitrary at this stage and its governing equation can be determined at higher orders. One needs to note that if $\hat{U}(Z,\omega)$ is real, then $C(\zeta)$ represents the chirp of $\hat{u}^{(0)}(\zeta, Z, \omega)$. Now, taking the inverse Fourier transform of Eq.(7.16) and using the convolution theorem of the Fourier transforms give

$$u^{(0)}(\zeta, Z, t) = \int_{-\infty}^{\infty} U(Z,t')p\,(C(\zeta), t-t')\,dt' \tag{7.17}$$

where

$$p(x,t) = \frac{1}{\sqrt{2\pi i x}}e^{\frac{it^2}{2x}} \tag{7.18}$$

Here, $\sqrt{x}$ is taken on a principal branch with $|\arg(x)| < \pi$. It needs to be noted that $\hat{p}(0,\omega) = 1$ so that $p(0,t) = \delta(t)$, the Dirac's delta function. Therefore, at ζ for which $C(\zeta) = 0$, gives $u^{(0)}(\zeta, Z, t) = U(Z, t)$ from Eq.(7.17). Also, $p(x,t)$ and $\hat{p}(x,\omega)$ are even in t and ω, respectively. Thus, the parity of $u^{(0)}(\cdot,\cdot,t)$ is determined by the parity of $U(\cdot,t)$; namely, if $U(Z,t)$ is even with respect to t, then so is $u^{(0)}(\zeta, Z, t)$ and vice versa. Now, taking the Fourier transform of the $O(1)$ equation, namely, Eq.(7.9), gives

$$-i\hat{L}[u^{(1)}] \equiv \frac{\partial \hat{u}^{(1)}}{\partial \zeta} + \frac{i\omega^2}{2}\Delta(\zeta)\hat{u}^{(1)} = i\hat{F}[u^{(0)}, v^{(0)}] \tag{7.19}$$

whose solution is

$$u^{(1)}(\zeta, Z, \omega) = \hat{p}(C(\zeta,\omega)\Big[\hat{U}^{(1)}(Z,\omega) + i\int_0^{\zeta} \hat{p}^*(C(\zeta'),\omega)\hat{F}[u^{(0)}, v^{(0)}](\zeta', Z, \omega)d\zeta'\Big] \tag{7.20}$$

where $\hat{U}^{(1)}$ needs to be determined at the next order in the perturbation expansion. To avoid secular terms, also known as resonances, in the perturbation expansion, the integrals must vanish over the dispersion map period z_a that corresponds to $\zeta = 1$. By Fredholm's Alternative (FA), [2, 3, 11–13] applied to Eq.(7.20) implies

$$\int_0^1 \hat{p}^*(C(\zeta),\omega)\hat{F}[u^{(0)}](\zeta, Z, \omega)d\zeta = 0 \tag{7.21}$$

This condition leads to the nonlinear evolution equation for the unknown function $\hat{U}(Z,\omega)$. To obtain this equation in the Fourier domain, the equation representing the FA, namely,Eq. (7.21), is first written down explicitly. Then $\hat{F}[u^{(0)}](\zeta, Z, \omega)$ is expressed as a Fourier transform of $F[u^{(0)}](\zeta, Z, t)$. Also, substitute $u^{(0)}$ from Eq.(7.17) to get the following integro-differential nonlinear evolution equations in the Fourier domain.

$$i\frac{\partial \hat{U}}{\partial Z} - \frac{\delta_a}{2}\omega^2\hat{U} + \int_{-\infty}^{\infty}\int_{-\infty}^{\infty} \hat{U}(Z,\omega+\omega_1)\hat{U}(Z,\omega+\omega_2) \cdot\hat{U}^*(Z,\omega+\omega_1+\omega_2)r(\omega_1\omega_2)\, d\omega_1 d\omega_2 = 0 \tag{7.22}$$

where the kernel $r(x)$ is given by

$$r(x) = \frac{1}{(2\pi)^2}\int_0^1 g(\zeta)e^{iC(\zeta)x}dx \tag{7.23}$$

By taking the inverse Fourier transform of Eq.(7.22), the following integro-differential nonlinear evolution equation in the temporal domain is obtained:

$$i\frac{\partial U}{\partial Z}+\frac{\delta_a}{2}\frac{\partial^2 U}{\partial t^2}+g(z)\int_{-\infty}^{\infty}\int_{-\infty}^{\infty}U(Z,t+t_1)U(Z,t+t_2)$$

$$\cdot U^*(Z,t+t_1+t_2)R(t_1,t_2)dt_1dt_2=0 \tag{7.24}$$

where the kernel is given by

$$R(t_1,t_2)=\int_{-\infty}^{\infty}\int_{-\infty}^{\infty}e^{i(\omega_1 t_1+\omega_2 t_2)}r(\omega_1\omega_2)d\omega_1 d\omega_2 \tag{7.25}$$

Equation (7.24) is known as the GTE due to the dispersion-managed solitons in optical fibers for a polarization-preserving fiber.

7.2.1 Special solutions

In this subsection, a traveling wave solution of the GTE that is given by Eq.(7.24), is being sought after. Now, the GTE equation is Galilean invariant, namely, if $U(Z,t)$ is a solution to Eq.(7.24), then $U(Z,t)e^{\frac{i}{2}(vt-\frac{v^2 Z}{2})}$ is also so for any real v. Thus, it is sufficient to look for solutions of the form $U(Z,t)=f(t)e^{\frac{i}{2}\lambda^2 Z}$ for real and even $f(t)$ and for nonnegative constant λ called the eigenvalue [3]. Then $F(\omega)$, the Fourier transform of $f(t)$, is also real and even. Thus, by virtue of Eq.(7.24), the following nonlinear integral equation for $F(\omega)$ is obtained:

$$(\lambda^2+\delta_a\omega^2)F(\omega)=2\int_{-\infty}^{\infty}\int_{-\infty}^{\infty}F(\omega+\omega_1)F(\omega+\omega_2)$$

$$\cdot F(\omega+\omega_1+\omega_2)r(\omega_1\omega_2)d\omega_1 d\omega_2 \tag{7.26}$$

It can be argued using scaling analysis that if $F_1(\omega)$ is a solution to Eq.(7.26) corresponding to the eigenvalue $\lambda=\lambda_1$ and the map strength $s=s_1$, then $F_2(\omega)=F_1(\omega/\nu)$ is also a solution to Eq.(7.26) corresponding to the eigenvalue $\lambda_2=\nu\lambda_1$ and the map strength $s_2=s_1/\nu^2$. Alternatively, if $F_1(\omega)$ is a solution corresponding to the average dispersion $\delta_a=1$ and $\lambda=\lambda_1$, then if $\delta_a>0$, $F(\omega)=\sqrt{\delta_a}F_1(\omega)$ is a solution corresponding to the average dispersion δ_a and $\lambda=\sqrt{\delta_a}\lambda_1$ for any fixed value of the map strength s. Also note that $f_2(t)=\nu f_1(\nu t)$ and thus the scaling parameter α is also directly proportional to the pulse energy and inversely proportional to the pulse width. So, $\|f_2\|^2=\nu\|f_1\|^2$ as in the case of a classical NLSE [3].

The following figures shows the direct numerical simulation of (7.26). Figure 7.1 is the shape of the stationary pulse in the Fourier domain for $s=1$ and $\lambda=4$, while Figure 7.2 represents the same pulse in the temporal domain.

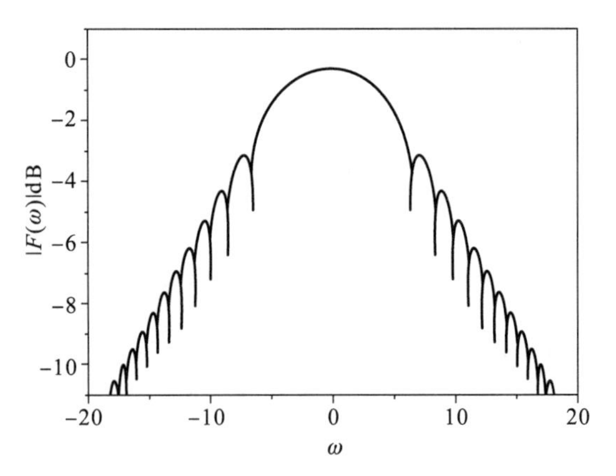

Fig. 7.1 Shape of stationary pulse in the Fourier domain for $s = 1, \lambda = 4$.

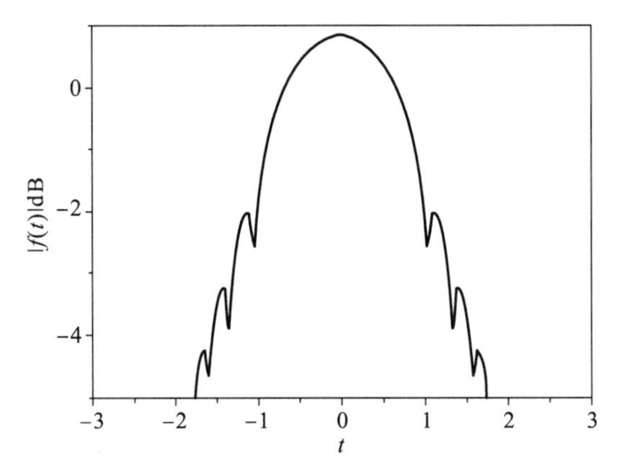

Fig. 7.2 Shape of stationary pulse in the temporal domain for $s = 1, \lambda = 4$.

7.3 Birefringent fibers

The material of this section is taken from the work of Biswas in 2001 [11]. The dimensionless form of the DM-VNLSE, that appears in the context of birefringent fibers, is going to be studied in this section by asymptotic analysis. The version of the DM-VNLSE that will be considered are Eqs.(4.3) and (4.4). They are

$$iu_z + \frac{D(z)}{2}u_{tt} + g(z)(|u|^2 + \alpha|v|^2)u = 0 \tag{7.27}$$

$$iv_z + \frac{D(z)}{2}v_{tt} + g(z)(|v|^2 + \alpha|u|^2)v = 0 \tag{7.28}$$

Equations (7.27) and (7.28) contains both large and rapidly varying terms. To obtain the asymptotic behavior, the fast and slow z scales, as in the previous section, are introduced. Now, expand the fields u and v in powers of z_a as follows [11]:

$$u(\zeta, Z, t) = u^{(0)}(\zeta, Z, t) + z_a u^{(1)}(\zeta, Z, t) + z_a^2 u^{(2)}(\zeta, Z, t) + \cdots \tag{7.29}$$

$$v(\zeta, Z, t) = v^{(0)}(\zeta, Z, t) + z_a v^{(1)}(\zeta, Z, t) + z_a^2 v^{(2)}(\zeta, Z, t) + \cdots \tag{7.30}$$

As before, decompose Eqs. (7.27) and (7.28) into a series of equations corresponding to the different powers of z_a. In general, at $O(z_a^{n-1})$,

$$F\left[u^{(n)}\right] = -P_n\left[u^{(0)}, v^{(0)}; u^{(1)}, v^{(1)}; \ldots; u^{(n-1)}, v^{(n-1)}\right] \tag{7.31}$$

and

$$F\left[v^{(n)}\right] = -P_n\left[v^{(0)}, u^{(0)}; v^{(1)}, u^{(1)}; \ldots; v^{(n-1)}, u^{(n-1)}\right] \tag{7.32}$$

where

$$F\left[u^{(n)}\right] = iu_\zeta^{(n)} + \frac{1}{2}\Delta(\zeta)u_{tt}^{(n)} \tag{7.33}$$

and

$$F\left[v^{(n)}\right] = iv_\zeta^{(n)} + \frac{1}{2}\Delta(\zeta)v_{tt}^{(n)} \tag{7.34}$$

with

$$P_0 = 0 \tag{7.35}$$

$$P_1\left[u^{(0)}, v^{(0)}\right] = iu_Z^{(0)} + \frac{1}{2}\delta_a u_{tt}^{(0)} + g(z)\left\{|u^{(0)}|^2 + \alpha|v^{(0)}|^2\right\}u^{(0)} \tag{7.36}$$

and

$$P_1\left[v^{(0)}, u^{(0)}\right] = iv_Z^{(0)} + \frac{1}{2}\delta_a v_{tt}^{(0)} + g(z)\left\{|v^{(0)}|^2 + \alpha|v^{(0)}|^2\right\}v^{(0)} \tag{7.37}$$

while

$$\begin{aligned} &P_2\left[u^{(0)}, v^{(0)}; u^{(1)}, v^{(1)}\right] \\ &= iu_Z^{(1)} + \frac{1}{2}\delta_a u_{tt}^{(1)} + g(z)\Big[2|u^{(0)}|^2 u^{(1)} + (u^{(0)})^2 (u^{(1)})^* \\ &\quad + \alpha\left\{2|v^{(0)}|^2 v^{(1)} + (v^{(0)})^2 (v^{(1)})^*\right\}\Big] \end{aligned} \tag{7.38}$$

and

$$\begin{aligned} &P_2\left[v^{(0)}, u^{(0)}; v^{(1)}, u^{(1)}\right] \\ &= iv_Z^{(1)} + \frac{1}{2}\delta_a v_{tt}^{(1)} + g(z)\Big[2|v^{(0)}|^2 v^{(1)} + (v^{(0)})^2 (v^{(1)})^* \\ &\quad + \alpha\left\{2|u^{(0)}|^2 u^{(1)} + (u^{(0)})^2 (u^{(1)})^*\right\}\Big] \end{aligned} \tag{7.39}$$

and so on. Again, the operators L and F are defined as

$$L[f] \equiv i\frac{\partial f}{\partial \zeta} + \frac{\Delta(\zeta)}{2}\frac{\partial^2 f}{\partial t^2}$$

$$F[f,h] \equiv i\frac{\partial f}{\partial z} + \frac{\delta_a}{2}\frac{\partial^2 f}{\partial t^2} + g(z)(|f|^2 + \alpha|h|^2)f$$

So, $O\left(1/z_a\right)$, yields

$$L[u^{(0)}] = 0 \tag{7.40}$$

and

$$L[v^{(0)}] = 0 \tag{7.41}$$

while at $O(1)$,

$$L[u^{(1)}] + F[u^{(0)}, v^{(0)}] = 0 \tag{7.42}$$

and

$$L[v^{(1)}] + F[v^{(0)}, u^{(0)}] = 0 \tag{7.43}$$

Taking the Fourier transforms of Eqs.(7.40) and (7.41), respectively, gives

$$-i\hat{L}[u^{(0)}] \equiv \frac{\partial \hat{u}^{(0)}}{\partial \zeta} + \frac{i\omega^2}{2}\Delta(\zeta)\hat{u}^{(0)} = 0 \tag{7.44}$$

and

$$-i\hat{L}[v^{(0)}] \equiv \frac{\partial \hat{v}^{(0)}}{\partial \zeta} + \frac{i\omega^2}{2}\Delta(\zeta)\hat{v}^{(0)} = 0 \tag{7.45}$$

whose respective solutions are

$$\hat{u}^{(0)}\left(\zeta, Z, \omega\right) = \hat{U}\left(Z, \omega\right)\hat{p}\left(C(\zeta), \omega\right) \tag{7.46}$$

and

$$\hat{v}^{(0)}\left(\zeta, Z, \omega\right) = \hat{V}\left(Z, \omega\right)\hat{p}\left(C(\zeta), \omega\right) \tag{7.47}$$

with $\hat{p}(x,\omega)$ and $C(\zeta)$ as before. Here, the integration constants $\hat{U}(Z,\omega)$ and $\hat{V}(Z,\omega)$ represent the slowly varying amplitudes of $u^{(0)}$ and $v^{(0)}$, respectively. The functions $\hat{U}(Z,\omega)$ and $\hat{V}(Z,\omega)$ are arbitrary at this stage and the equations governing them will be determined at higher orders. Just as in the case of polarization-preserving fibers, if $\hat{V}(Z,\omega)$ is real, then $C(\zeta)$ represents the chirp of $\hat{v}^{(0)}(\zeta, Z, \omega)$. Now, taking the inverse Fourier transforms of Eqs.(7.44) and (7.45) and using the convolution theorem of Fourier transforms implies

$$u^{(0)}(\zeta, Z, t) = \int_{-\infty}^{\infty} U(Z, t')p\left(C(\zeta), t - t'\right)dt' \tag{7.48}$$

and

$$v^{(0)}(\zeta, Z, t) = \int_{-\infty}^{\infty} V(Z, t')p\left(C(\zeta), t - t'\right)dt' \tag{7.49}$$

where $p(x,t)$ is defined in Eq.(7.18). Here again, the parity of $v^{(0)}(\cdot,\cdot,t)$ is determined by the parity of $V(\cdot,t)$; namely, if $V(Z,t)$ is even with respect to t, then so is $v^{(0)}(\zeta,Z,t)$.

Now, taking the Fourier transform of the $O(1)$ equations, namely, Eqs.(7.42) and (7.43), respectively, gives

$$-i\hat{L}[u^{(1)}] \equiv \frac{\partial \hat{u}^{(1)}}{\partial \zeta} + \frac{i\omega^2}{2}\Delta(\zeta)\hat{u}^{(1)} = i\hat{F}[u^{(0)}, v^{(0)}] \tag{7.50}$$

and

$$-i\hat{L}[v^{(1)}] \equiv \frac{\partial \hat{v}^{(1)}}{\partial \zeta} + \frac{i\omega^2}{2}\Delta(\zeta)\hat{v}^{(1)} = i\hat{F}[v^{(0)}, u^{(0)}] \tag{7.51}$$

whose solutions are, respectively,

$$u^{(1)}(\zeta, Z, \omega) = \hat{p}(C(\zeta,\omega)\Big[\hat{U}^{(1)}(Z,\omega) + i\int_0^\zeta \hat{p}^*(C(\zeta'),\omega)\hat{F}[u^{(0)}, v^{(0)}](\zeta', Z, \omega)d\zeta'\Big] \tag{7.52}$$

and

$$v^{(1)}(\zeta, Z, \omega) = \hat{p}(C(\zeta,\omega)\Big[\hat{V}^{(1)}(Z,\omega) + i\int_0^\zeta \hat{p}^*(C(\zeta'),\omega)\hat{F}[v^{(0)}, u^{(0)}](\zeta', Z, \omega)d\zeta'\Big] \tag{7.53}$$

where $\hat{U}^{(1)}$ and $\hat{V}^{(1)}$ are to be determined at the next order in the perturbation expansion. Again, to avoid secular terms use the FA to Eqs.(7.52) and (7.53) that respectively yield

$$\int_0^1 \hat{p}^*(C(\zeta),\omega)\hat{F}[u^{(0)}, v^{(0)}](\zeta, Z, \omega)d\zeta = 0 \tag{7.54}$$

and

$$\int_0^1 \hat{p}^*(C(\zeta),\omega)\hat{F}[v^{(0)}, u^{(0)}](\zeta, Z, \omega)d\zeta = 0 \tag{7.55}$$

These conditions lead to the nonlinear evolution equation for the unknown functions $\hat{U}(Z,\omega)$ and $\hat{V}(Z,\omega)$. To obtain these equations in the Fourier domain, write explicitly the equations representing the FA, namely, Eqs.(7.54) and (7.55). Then express $\hat{F}[u^{(0)}, v^{(0)}](\zeta, Z, \omega)$ and $\hat{F}[v^{(0)}, u^{(0)}](\zeta, Z, \omega)$ as Fourier transforms of $F[u^{(0)}, v^{(0)}](\zeta, Z, t)$ and $F[v^{(0)}, u^{(0)}](\zeta, Z, t)$, respectively. Also substitute $u^{(0)}$ and $v^{(0)}$ from Eqs.(7.46) and (7.47), respectively, to get the following coupled integro-differential nonlinear evolution equations in the Fourier domain [3]:

$$
\begin{aligned}
&i\frac{\partial \hat{U}}{\partial Z} - \frac{\delta_a}{2}\omega^2 \hat{U} + \int_{-\infty}^{\infty}\int_{-\infty}^{\infty} r(\omega_1\omega_2)\hat{U}(Z,\omega_1+\omega_2) \\
&\cdot\Big[\hat{U}(Z,\omega+\omega_1)\hat{U}^*(Z,\omega+\omega_1+\omega_2) \\
&+ \alpha\hat{V}(Z,\omega+\omega_1)\hat{V}^*(Z,\omega+\omega_1+\omega_2)\Big] d\omega_1 d\omega_2 = 0
\end{aligned} \tag{7.56}
$$

and

$$
\begin{aligned}
&i\frac{\partial \hat{V}}{\partial Z} - \frac{\delta_a}{2}\omega^2 \hat{V} + \int_{-\infty}^{\infty}\int_{-\infty}^{\infty} r(\omega_1\omega_2)\hat{V}(Z,\omega_1+\omega_2) \\
&\cdot\Big[\hat{V}(Z,\omega+\omega_1)\hat{V}^*(Z,\omega+\omega_1+\omega_2) \\
&+ \alpha\hat{U}(Z,\omega+\omega_1)\hat{U}^*(Z,\omega+\omega_1+\omega_2)\Big] d\omega_1 d\omega_2 = 0
\end{aligned} \tag{7.57}
$$

By taking the inverse Fourier transform of Eqs.(7.56) and (7.57), we get the following coupled integro-differential nonlinear evolution equations in the time domain:

$$
\begin{aligned}
&i\frac{\partial U}{\partial Z} + \frac{\delta_a}{2}\frac{\partial^2 U}{\partial t^2} + g(z)\int_{-\infty}^{\infty}\int_{-\infty}^{\infty} R(t_1,t_2)U(Z,t_1+t_2) \\
&\cdot[U(Z,t+t_1)U^*(Z,t+t_1+t_2) \\
&+ \alpha V(Z,t+t_1)V^*(Z,t+t_1+t_2)]dt_1 dt_2 = 0
\end{aligned} \tag{7.58}
$$

and

$$
\begin{aligned}
&i\frac{\partial V}{\partial Z} + \frac{\delta_a}{2}\frac{\partial^2 V}{\partial t^2} + g(z)\int_{-\infty}^{\infty}\int_{-\infty}^{\infty} R(t_1,t_2)V(Z,t_1+t_2) \\
&\cdot[V(Z,t+t_1)V^*(Z,t+t_1+t_2) \\
&+ \alpha U(Z,t+t_1)U^*(Z,t+t_1+t_2)]dt_1 dt_2 = 0
\end{aligned} \tag{7.59}
$$

The kernels in Eqs.(7.58) and (7.59) stay the same as in the case of polarization preserving fibers. Equations (7.58) and (7.59) are the GTE for DM solitons in birefringent fibers.

7.4 DWDM system

The material of this section is also taken from the work of Biswas that first appeared in 2003 [12, 13]. The NLSE for DWDM system is studied in Chapter 5. It is this equation that will be studied in this section by asymptotic analysis. Thus, to start off, the governing equation is

$$iu_{l,z} + \frac{D(z)}{2} u_{l,tt} + g(z) \left\{ |u_l|^2 + \sum_{m \neq l}^{N} \alpha_{lm} |u_m|^2 \right\} u_l = 0 \tag{7.60}$$

where $1 \leq l \leq N$. This represents multi-channel WDM transmission of co-propagating wave envelopes in a nonlinear optical fiber, including the XPM effect. Just to recall, α_{lm} are known as the XPM coefficients. Once again, Eq. (7.60) contains both large and rapidly varying terms. So, to obtain the asymptotic behavior introduce the fast and slow z scales, as usual, and expand the fields u_l in powers of z_a as [12,13]

$$u_l(\zeta, Z, t) = u_l^{(0)}(\zeta, Z, t) + z_a u_l^{(1)}(\zeta, Z, t) + z_a^2 u_l^{(2)}(\zeta, Z, t) + \cdots \tag{7.61}$$

Subsequently, decompose Eq. (7.60) into a series of equations corresponding to the different powers of z_a. In general, at $O(z_a^{n-1})$,

$$F\left[u_l^{(n)}, u_m^{(n)}\right] = -P_n\left[u_l^{(0)}, u_m^{(0)}; u_l^{(1)}, u_m^{(1)}; \ldots; u_l^{(n-1)}, u_m^{(n-1)}\right] \tag{7.62}$$

where

$$F\left[u_l^{(n)}\right] = iu_{l,\zeta}^{(n)} + \frac{1}{2}\Delta(\zeta) u_{l,tt}^{(n)} \tag{7.63}$$

and

$$P_0 = 0 \tag{7.64}$$

$$P_1\left[u_l^{(0)}, u_m^{(0)}\right] = iu_Z^{(0)} + \frac{1}{2}\delta_a u_{tt}^{(0)} + g(z)|u^{(0)}|^2 u^{(0)} + g(z) \sum_{m \neq l}^{N} \alpha_{lm} |u_m^{(0)}|^2 u_m^{(0)} \tag{7.65}$$

while

$$\begin{aligned} P_2&\left[u_l^{(0)}, u_m^{(0)}; u_l^{(1)}, u_m^{(1)}\right] \\ &= iu_Z^{(1)} + \frac{1}{2}\delta_a u_{tt}^{(1)} + g(z)\left\{(u_l^{(0)})^2 (u_l^{(1)})^* + 2|u_l^{(0)}|^2 u_l^{(1)}\right\} \\ &\quad + g(z) \sum_{m \neq l}^{N} \alpha_{lm} \left\{|u_m^{(0)}|^2 (u_l^{(1)})^* + u_l^{(0)}(u_m^{(0)}(u_m^{(1)})^* + u_m^{(1)}(u_m^{(0)})^*)\right\} \end{aligned} \tag{7.66}$$

and so on. The notation

$$u_{l,\zeta}^{(n)} = \frac{\partial u_l^{(n)}}{\partial \zeta}$$

and

$$u_{l,tt}^{(n)} = \frac{\partial^2 u_l^{(n)}}{\partial t^2}$$

was used. Again, with the operators L and F defined as

$$L[f_l] \equiv i\frac{\partial f_l}{\partial \zeta} + \frac{\Delta(\zeta)}{2}\frac{\partial^2 f_l}{\partial t^2}$$

$$F[f_l, h_m] \equiv i\frac{\partial f_l}{\partial z} + \frac{\delta_a}{2}\frac{\partial^2 f_l}{\partial t^2} + g(z)\left(|f_l|^2 + \sum_{m \neq l}^{N} \alpha_{lm}|h_m|^2\right) f_l$$

at $O(1/z_a)$ gives

$$L[u_l^{(0)}] = 0 \tag{7.67}$$

while at $O(1)$ gives

$$L[u_l^{(1)}] + F[u_l^{(0)}, u_m^{(0)}] = 0 \tag{7.68}$$

Taking the Fourier transform of (7.67) gives

$$-i\hat{L}[u_l^{(0)}] \equiv \frac{\partial \hat{u_l}^{(0)}}{\partial \zeta} + \frac{i\omega^2}{2}\Delta(\zeta)\hat{u_l}^{(0)} = 0 \tag{7.69}$$

whose solution is

$$\hat{u_l}^{(0)}(\zeta, Z, \omega) = \hat{U_l}(Z, \omega)\, \hat{p}\left(C(\zeta), \omega\right) \tag{7.70}$$

Once again, the integration constant $\hat{U_l}(Z, \omega)$ represents the slowly varying amplitude of $u_l^{(0)}$, while $\hat{p}\left(C(\zeta), \omega\right)$ contains the fast periodic oscillations due to large local values of the dispersion. The function $\hat{U_l}(Z, \omega)$ is arbitrary at this stage and the equation governing them needs to be determined at higher orders. Now, taking the inverse Fourier transform of Eq.(7.70) and using the convolution theorem of the Fourier transforms give

$$u_l^{(0)}(\zeta, Z, t) = \int_{-\infty}^{\infty} U_l(Z, t')p\left(C(\zeta), t - t'\right) dt' \tag{7.71}$$

where $p(x, t)$ is defined in Eq.(7.18). Now, taking the Fourier transform of the $O(1)$ equation, namely, Eq.(7.68), gives

$$-i\hat{L}[u_l^{(1)}] \equiv \frac{\partial \hat{u_l}^{(1)}}{\partial \zeta} + \frac{i\omega^2}{2}\Delta(\zeta)\hat{u_l}^{(1)} = i\hat{F}[u_l^{(0)}, u_m^{(0)}] \tag{7.72}$$

whose solution is

$$u_l^{(1)}(\zeta, Z, \omega) = \hat{p}(C(\zeta, \omega)\Big[\hat{U_l}^{(1)}(Z, \omega)$$

$$+ i\int_0^{\zeta} \hat{p}^*(C(\zeta'), \omega)\hat{F}[u_l^{(0)}, u_m^{(0)}](\zeta', Z, \omega)d\zeta'\Big] \tag{7.73}$$

where $\hat{U_l}^{(1)}$ is to be determined at the next order in the perturbation expansion. To kill secular terms, apply FA to Eq.(7.73), thus resulting in

$$\int_0^1 \hat{p}^*(C(\zeta),\omega)\hat{F}[u_l^{(0)},u_m^{(0)}](\zeta,Z,\omega)d\zeta = 0 \tag{7.74}$$

This condition leads to the nonlinear evolution equation for the unknown functions $\hat{U_l}(Z,\omega)$. To obtain this equation in the Fourier domain, write explicitly the equations representing the FA, namely, Eq.(7.74). Then express $\hat{F}[u_l^{(0)},u_m^{(0)}](\zeta,Z,\omega)$ as a Fourier transform of $F[u_l^{(0)},u_m^{(0)}](\zeta,Z,t)$. Also substitute $u_l^{(0)}$ from Eq.(7.71) to get the following coupled integro-differential nonlinear evolution equations in the Fourier domain:

$$\begin{aligned} &i\frac{\partial \hat{U_l}}{\partial Z} - \frac{\delta_a}{2}\omega^2\hat{U_l} + \int_{-\infty}^{\infty}\int_{-\infty}^{\infty} r(\omega_1\omega_2)\hat{U_l}(Z,\omega_1+\omega_2) \\ &\cdot\Bigg[\hat{U_l}(Z,\omega+\omega_1)\hat{U_l}^*(Z,\omega+\omega_1+\omega_2) \\ &+\sum_{l\neq m}^{N}\alpha_{lm}\hat{U_m}(Z,\omega+\omega_1)\hat{U_m}^*(Z,\omega+\omega_1+\omega_2)\Bigg]d\omega_1 d\omega_2 = 0 \end{aligned} \tag{7.75}$$

By taking the inverse Fourier transform of Eq.(7.75) the following coupled integro-differential nonlinear evolution equations in the time domain is obtained:

$$\begin{aligned} &i\frac{\partial U_l}{\partial Z} + \frac{\delta_a}{2}\frac{\partial^2 U_l}{\partial t^2} + g(z)\int_{-\infty}^{\infty}\int_{-\infty}^{\infty} R(t_1,t_2)U_l(Z,t_1+t_2) \\ &\cdot\Bigg[U_l(Z,t+t_1)U_l^*(Z,t+t_1+t_2) \\ &+\sum_{l\neq m}^{N}\alpha_{lm}U_m(Z,t+t_1)U_m^*(Z,t+t_1+t_2)\Bigg]dt_1 dt_2 = 0 \end{aligned} \tag{7.76}$$

where $1 \leq l \leq N$ and the kernels stay the same as in the case of polarization preserving fibers. Equation (7.76) is the GTE for DWDM systems.

7.5 Properties of the kernel

The GTE equations for different types of fibers are the fundamental equations that govern the evolution of optical pulses for a strong dispersion-managed soliton system corresponding to the frequency and time domain, respectively. In the GTE, all the fast variations and large quantities are removed and therefore contain only slowly varying quantities of order one. These equations are not limited to the case $\delta_a > 0$, however, they are also applicable to the case of pulse dynamics with zero or normal values of average dispersion. If the fiber dispersion is constant, namely, if $\Delta(\zeta) = 0$ and $C_0 = 0$, then $C(\zeta) = 0$ and so $r(x) = 1/(2\pi)^2$ and thus, $R(t_1, t_2) = \delta(t_1)\delta(t_2)$, a two-dimensional Dirac's delta function. The kernel $r(x)$ will now be studied in the following two cases.

7.5.1 Lossless Case

For the lossless case, namely, when $g(\zeta) = 1$, note that, the kernel $r(x)$ for a two-step map defined in Eq.(2.34) takes a very simple form, namely,

$$r(x) = \frac{1}{(2\pi)^2} \frac{\sin(sx)}{sx} \tag{7.77}$$

so that

$$R(t_1, t_2) = \frac{1}{2\pi} \frac{\mathrm{ci}(\frac{t_1 t_2}{s})}{|s|} \tag{7.78}$$

where the cosine integral $\mathrm{ci}(x)$ is defined as

$$\mathrm{ci}(x) = \int_1^\infty \frac{\cos(xy)}{y} dy$$

Since $|r(x)| \leq 1/2\pi$, the strength of the coupling between different frequencies that measures the effective nonlinearity of the system is always less than that of the ordinary vector NLSE.

7.5.2 Lossy Case

For the lossy case, namely, when $g(\zeta) \neq 1$, the kernel $r(x)$ depends on the relative position of the amplifier with respect to the dispersion map. Consider the two-step map given by $\Delta(\zeta)$ in Eq.(2.34) and define ζ_a to represent the position of the amplifier within the dispersion map. So $|\zeta_a| < 1/2$ and $\zeta_a = 0$ means that the amplifier is placed at the mid point of the anomalous fiber segment. The function $g(\zeta)$ given in Eq.(2.42) can then be written as

$$g(\zeta) = \frac{2Ge^{2G}}{\sinh(2G)} e^{-4G(\zeta - n\zeta_a)} \tag{7.79}$$

for $\zeta_a + n \leq \zeta < \zeta_a + n + 1$, where $G = \Gamma z_a/2$. The kernel $r(x)$ in the lossy case is computed in a similar method as in the lossless case. If $|\zeta_a| < \theta/2$, namely, the amplifier is located in the anomalous fiber segment, the resulting expression for kernel is

$$
\begin{aligned}
r(x) = \frac{1}{(2\pi)^2} & \frac{Ge^{iC_0 x}}{(sx + 2iG\theta)(sx - 2iG(1-\theta))} \\
& \cdot \left[e^{G(4\zeta_a - 2\theta + 1)} sx \frac{\sin(sx - 2iG(1-\theta))}{\sinh(2G)} \right. \\
& \left. + i\theta e^{i(4\zeta_a - 2\theta + 1)\frac{sx}{2\theta}} (sx - 2iG(1-\theta)) \right]
\end{aligned}
\tag{7.80}
$$

In Eq.(7.80), unlike the lossless case, the kernel $r(x)$ is complex and is explicitly dependent on the parameters θ, Γ, z_a and ζ_a in a nontrivial way. However, one still obtains

$$\lim_{s \to 0} r(x) = \frac{1}{(2\pi)^2} \tag{7.81}$$

and moreover,

$$\lim_{G \to 0} r(x) = \frac{1}{(2\pi)^2} \frac{\sin(sx)}{sx} \tag{7.82}$$

which means that as $z_a \to 0$, Eq.(7.77) is recovered. For the particular case $\theta = 1/2$, $\zeta_a = 0$ and $C_0 = 0$ which corresponds to fiber segments of equal length with amplifiers placed at the middle of the anomalous fiber segment, the kernel modifies to

$$r(x) = \frac{1}{(2\pi)^2} \frac{G}{x^2 s^2 + G^2} \left[sx \frac{\sin(sx)}{\sinh G} + isx \left\{ 1 - \frac{\cos(sx)}{\cosh G} \right\} + G \right] \tag{7.83}$$

References

1. F. Abdullaev, S. Darmanyan & P. Khabibullaev. *Optical Solitons.* Springer, New York, NY. USA. (1993).
2. M. J. Ablowitz & H. Segur. *Solitons and the Inverse Scattering Transform.* SIAM. Philadelphia, PA. USA. (1981).
3. M. J. Ablowitz & G. Biondini. "Multiscale pulse dynamics in communication systems with strong dispersion management". *Optics Letters.* Vol 23, No 21, 1668-1670. (1998).
4. M. J. Ablowitz & T. Hirooka. "Nonlinear effects in quasi-linear dispersion-managed systems". *IEEE Photonics Technology Letters.* Vol 13, No 10, 1082-1084. (2001).
5. M. J. Ablowitz, G. Biondini & E. S. Olson. "Incomplete collisions of wavelength-division multiplexed dispersion-managed solitons". *Journal of Optical Society of America B.* Vol 18, No 3, 577-583. (2001).
6. M. J. Ablowitz & T. Hirooka. "Nonlinear effects in quasilinear dispersion-managed pulse transmission". *IEEE Journal of Photonics Technology Letters.* Vol 26, 1846-1848. (2001).

7. M. J. Ablowitz & T. Hirooka. "Intrachannel pulse interactions and timing shifts in strongly dispersion-managed transmission systems". *Optics Letters*. Vol 26, No 23, 1846-1848. (2001).
8. M. J. Ablowitz & T. Hirooka. "Intrachannel pulse interactions in dispersion-managed transmission systems: energy transfer". *Optics Letters*. Vol 27, No 3, 203-205. (2002).
9. M. J. Ablowitz & T. Hirooka. "Managing nonlinearity in strongly dispersion-managed optical pulse transmission". *Journal of Optical Society of America B*. Vol 19, No 3, 425-439. (2002).
10. M. J. Ablowitz, G. Biondini, A. Biswas, A. Docherty, T. Hirooka & S. Chakravarty. "Collision-induced timing shifts in dispersion-managed soliton systems". *Optics Letters*. Vol 27, 318-320. (2002).
11. A. Biswas. "Dispersion-managed vector solitons in optical fibers". *Fiber and Integrated Optics*. Vol 20, No 5, 503-515. (2001).
12. A. Biswas. "Gabitov-Turitsyn equation for solitons in multiple channels". *Journal of Electromagnetic Waves and Applications*. Vol 17, No 11, 1539-1560. (2003).
13. A. Biswas. "Gabitov-Turitsyn equation for solitons in optical fibers". *Journal of Nonlinear Optical Physics and Materials*. Vol 12, No 1, 17-37. (2003).
14. D. Breuer, F. Kuppers, A. Mattheus, E. G. Shapiro, I. Gabitov & S. K. Turitsyn. "Symmetrical dispersion compensation for standard monomode fiber based communication systems with large amplifier spacing". *Optics Letters*. Vol 22, No 13, 982-984. (1997).
15. D. Breuer, K. Jürgensen, F. Küppers, A. Mattheus, I. Gabitov & S. K. Turitsyn. "Optimal schemes for dispersion compensation of standard monomode fiber based links". *Optics Communications*. Vol 40, No 1-3, 15-18. (1997).
16. I. R. Gabitov & S. K. Turitsyn. "Average pulse dynamics in a cascaded transmission system with passive dispersion compensation". *Optics Letters*. Vol 21, No 5, 327-329. (1996).
17. I. R. Gabitov & S. K. Turitsyn. "Breathing solitons in optical fiber links". *JETP Letters*. Vol 63, No 10, 861-866. (1996).
18. I. R. Gabitov, E. G. Shapiro & S. K. Turitsyn. "Asymptotic breathing pulse in optical transmission systems with dispersion compensation". *Physical Review E*. Vol 55, No 3, 3624-3633. (1997).
19. I. R. Gabitov & P. M. Lushnikov. "Nonlinearity management in a dispersion compensation system". *Optics Letters*. Vol 27, No 2, 113-115. (2002).
20. I. R. Gabitov, R. Indik, L. Mollenauer, M. Shkarayev, M. Stepanov & P. M. Lushnikov. "Twin families of bisolitons in dispersion-managed systems". *Optics Letters*. Vol 32, No 6, 605-607. (2007).
21. A-H. Guan & Y-H. Wang. "Experimental study of interband and intraband crosstalk in WDM networks". *Optoelectronics Letters*. Vol 4, No 1, 42-44. (2008).
22. T. Hirooka & A. Hasegawa. "Chirped soliton interaction in strongly dispersion-managed wavelength-division-multiplexing systems". *Optics Letters*. Vol 23, No 10, 768-770. (1998).
23. T. Hirooka & S. Wabnitz. "Stabilization of dispersion-managed soliton transmission by nonlinear gain". *Electronics Letters*. Vol 35, No 8, 655-657. (1999).
24. T. Hirooka & S. Wabnitz. "Nonlinear gain control of dispersion-managed soliton amplitude and collisions". *Optical Fiber Technology*. Vol 6, No 2, 109-121. (2000).
25. V. E. Zakharov & S. Wabnitz. *Optical Solitons: Theoretical Challenges and Industrial Perspectives*. Springer, Heidelberg. DE. (1999).

Chapter 8
Quasi-linear Pulses

8.1 Introduction

In quasi-linear or low-powered systems, strong dispersion-management (DM), which uses multiple section of fibers that alternates positive and negative group-velocity dispersions, is used to compensate the fiber dispersion, manage fiber nonlinearity and suppress the inter-channel crosstalk. The difference between the DM soliton transmission and the quasi-linear pulses is that in the former, nonlinearity balances dispersion, while in the latter, nonlinearity is managed. In the past few years, there has been quite a few theoretical, numerical and experimental studies that were conducted in regards to the quasi-linear pulse transmission. In this chapter, the mathematical theory of the quasi-linear pulse transmission through optical fibers will be studied in details. The GTE is considered in all these three cases and the asymptotic analysis is carried out. The significant reduction in the effective nonlinearity for large map strength is quantified.

The study of quasi-linear pulses is split into three sections corresponding to the polarization preserving fibers, birefringent fibers as well as the DWDM systems. The mathematical theory of quasi-linear pulses was first studied by Ablowitz et al. in 2001 [6, 11] for polarization-preserving fibers. Later, in 2004, this study was extended to the case of birefringent fibers and DWDM systems by Biswas [19]. The first section of this chapter is therefore taken from Ablowitz's works that first appeared in 2001, while the second and the third sections of this chapter are taken from Biswas's work that appeared in 2004 [19].

8.2 Polarization-preserving fibers

Recall the GTE in the Fourier domain, for pulses in polarization preserving fibers is given by

$$i\frac{\partial \hat{U}}{\partial z} - \frac{\delta_a}{2}\omega^2\hat{U} + \int_{-\infty}^{\infty}\int_{-\infty}^{\infty} \hat{U}(z,\omega+\omega_1)\hat{U}(z,\omega+\omega_2)$$

$$\cdot\hat{U}^*(z,\omega+\omega_1+\omega_2)r(\omega_1\omega_2)d\omega_1 d\omega_2 = 0 \tag{8.1}$$

where the kernel $r(x)$ is given by

$$r(x;s) = \frac{1}{(2\pi)^2}\int_0^1 g(\zeta)e^{iC(\zeta)x}dx \tag{8.2}$$

while the GTE in temporal domain is given by

$$i\frac{\partial U}{\partial z} + \frac{\delta_a}{2}\frac{\partial^2 U}{\partial t^2} + g(z)\int_{-\infty}^{\infty}\int_{-\infty}^{\infty} U(z,t+t_1)U(z,t+t_2)$$

$$\cdot U^*(z,t+t_1+t_2)R(t_1,t_2)dt_1 dt_2 = 0 \tag{8.3}$$

where the kernel $R(t_1,t_2)$ is

$$R(t_1,t_2;s) = \int_{-\infty}^{\infty}\int_{-\infty}^{\infty} e^{i(\omega_1 t_1+\omega_2 t_2)}r(\omega_1\omega_2)d\omega_1 d\omega_2 \tag{8.4}$$

In Eq.(8.3), $U(z,t)$ is the slowly varying amplitude of $u(z,t)$ at the leading order. This equation is the universal asymptotic equation that governs the evolution of the amplitude of an optical pulse for a dispersion-managed system that is given by Eq.(2.41). Here in Eq.(8.3), all fast and large variations are removed. Recall here that Eq.(8.3) is equally applicable to the case of pulse dynamics for a zero or normal value of average dispersions.

In this section, the details of the nonlinear terms of the GTE, for large s will be analyzed. This analysis will also explain the difference between the quasi-linear transmission and soliton propagation. This study will be split into two subsections that deal with the lossless and the lossy cases, respectively.

8.2.1 Lossless system

In a lossless system, namely, $g(z) = 1$, the kernels given by Eqs.(8.2) and (8.4) respectively modify to

$$r(x;s) = \frac{1}{(2\pi)^2}\frac{\sin(sx)}{sx} \tag{8.5}$$

and

$$R(t_1, t_2; s) = \frac{1}{2\pi} \frac{\mathrm{ci}(\frac{t_1 t_2}{s})}{|s|} \tag{8.6}$$

where the cosine integral $\mathrm{ci}(x)$ is defined as

$$\mathrm{ci}(x) = \int_1^{\infty} \frac{\cos(xy)}{y} dy \tag{8.7}$$

Note that from Eq.(8.5), one can write

$$\lim_{s \to \infty} r(x; s) = 0 \tag{8.8}$$

thus showing that, for large map strength, one obtains the linear evolution equations. Assuming that $\hat{U}(z, \omega)$ depends weakly on s, the following asymptotic expansions of the nonlinear terms from the GT equation are obtained:

$$i\frac{\partial U}{\partial Z} + \frac{\delta_a}{2}\frac{\partial^2 U}{\partial t^2} + \frac{1}{2\pi s}\left[(\log s - \gamma) J_1(z, t) - J_2(z, t)\right] = 0 \tag{8.9}$$

where

$$J_1(z, t) = \int_{-\infty}^{\infty} \int_{-\infty}^{\infty} U(z, t + t_1) U(z, t + t_2) \cdot U^*(z, t + t_1 + t_2)\, dt_1 dt_2 \tag{8.10}$$

$$J_2(z, t) = \int_{-\infty}^{\infty} \int_{-\infty}^{\infty} \log|t_1 t_2| U(z, t + t_1) U(z, t + t_2) \cdot U^*(z, t + t_1 + t_2) dt_1 dt_2 \tag{8.11}$$

and

$$\gamma = \lim_{n \to \infty} \left[1 + \frac{1}{2} + \frac{1}{3} + \frac{1}{4} + \cdots + \frac{1}{n} - \log n\right] = 0.57721\ldots \tag{8.12}$$

is the Euler's constant. In the Fourier domain, Eq.(8.9) transforms to [3, 8]

$$i\frac{\partial \hat{U}}{\partial z} - \frac{\delta_a \omega^2}{2}\hat{U} + \frac{1}{2\pi s}\left[(\log s - \gamma)\hat{J}_1(z, \omega) - \hat{J}_2(z, \omega)\right] = 0 \tag{8.13}$$

where

$$\hat{J}_1(z, \omega) = \left|\hat{U}(z, \omega)\right|^2 \hat{U}(z, \omega) \tag{8.14}$$

$$\hat{J}_2(z, \omega) = \frac{1}{\pi}\hat{U}(z, \omega) \int_{-\infty}^{\infty} \left|\hat{U}(z, \omega')\right|^2 h(\omega' - \omega)\, d\omega' \tag{8.15}$$

and

$$h(\omega) = \int_{-\infty}^{\infty} \log|t| e^{-i\omega t} dt \tag{8.16}$$

These asymptotic results can be obtained by the stationary phase method applied to Eq.(2.41). They can also be derived directly from Eq.(8.1) and using the asymptotic expansion of the cosine integral, namely,

$$\mathrm{ci}(x) \sim -\gamma - \ln x + O(x) \tag{8.17}$$

as $x \to 0$. Now, neglecting the $O(1/s^2)$ term, one can see that the spectral intensity given by $|\hat{U}(z,\omega)|^2$ is preserved during the pulse propagation. Now, from (8.13),

$$\frac{\partial}{\partial z}|\hat{U}|^2 = 0 \tag{8.18}$$

Also, the solution of (8.13) is

$$\hat{U}(z,\omega) = \hat{U}(z,0)\exp\left\{ - i\frac{\delta_a\omega^2}{2}z + i\psi\left[|\hat{U}(0,\omega)|^2\right]z\right\} \tag{8.19}$$

where

$$\psi\left[|\hat{U}(z,\omega)|^2\right] = \frac{1}{2\pi s}\left[(\log s - \gamma)|\hat{U}(z,\omega)|^2 \right.$$
$$\left. -\frac{1}{\pi}\int_{-\infty}^{\infty}|\hat{U}(z,\omega)|^2 h(\omega' - \omega)d\omega'\right] \tag{8.20}$$

The linear phase shift $\exp\left(-i\delta_a\omega^2 z/2\right)$ can be corrected by the pre-transmission or the post-transmission compensation. After this, the linear phase is removed. However, if the system has a small value of the path averaged dispersion, namely, if $\delta_a \ll 1$, the average dynamics of the quasi-linear pulse transmission through an optical fiber is characterized only by the nonlinear phase shift

$$\phi_{NL}(z,\omega) = \psi\left[|\hat{U}(0,\omega)|^2\right]z.$$

The nonlinear chirp, for small values of the map strength (s), can induce pulse broadening. However, there is a simple way to compensate for the nonlinear chirp by expanding $\phi_{NL}(z,\omega)$ in a Taylor series with respect to ω:

$$\phi_{NL}(z,\omega) = \phi_{NL}(z,0) + \frac{\omega^2}{2!}\phi''_{NL}(z,0) + \frac{\omega^2}{4!}\phi''''_{NL}(z,0) + \cdots$$

with $|\hat{U}(0,\omega)|^2$ assumed to be even.

Also, note that large values of s reduce the effective nonlinearity. Therefore, a pulse in a polarization preserved fiber will be able to propagate for much longer distances before being distorted by nonlinearity, as opposed to a pulse, with the same energy, in a system with constant dispersion. Thus, strong DM allows the propagation of a stationary pulse in a quasi-linear regime, with energies comparable to that of classical solitons, but at the same time, much lower than the energy required for the formation of a stable DM soliton at $\delta_a \sim 0$ for the same value of s owing to the energy enhancement of DM solitons. Figure 8.1 shows a quasi-linear Gaussian pulse in the lossless case.

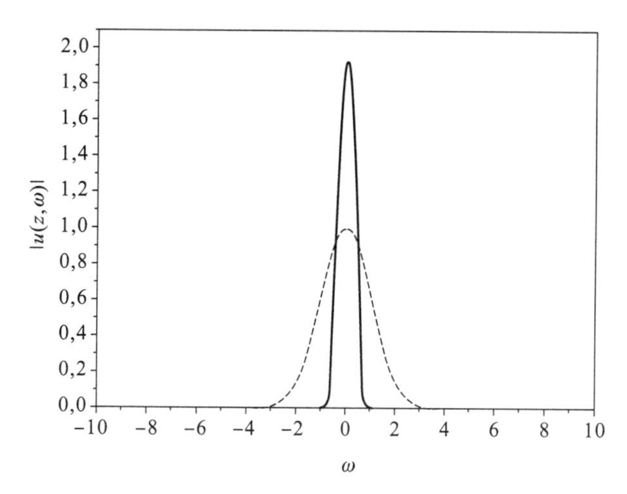

Fig. 8.1 Shape of the quasi-linear Gaussian pulse after propagation of $z = 20$ and the initial profile (dotted curve) for $s = 100$, $\langle d \rangle = 0$, and $G = 0$ in frequency domain.

8.2.2 Lossy system

For the lossy case, namely, when $g(\zeta) \neq 1$, the kernel $r(x; s)$ depends on the relative position of the amplifier with respect to the dispersion map. Consider the two-step map given by $\Delta(\zeta)$ in Eq.(2.34) and define ζ_a to represent the position of the amplifier within the dispersion map. So $|\zeta_a| < 1/2$ and $\zeta_a = 0$ means that the amplifier is placed at the mid point of the anomalous fiber segments. The function $g(\zeta)$ given by Eq.(2.42) can then be written as

$$g(\zeta) = \frac{2Ge^{2G}}{\sinh(2G)} e^{-4G(\zeta - n - \zeta_a)} \tag{8.21}$$

for $\zeta_a + n \leq \zeta < \zeta_a + n + 1$, where $G = \Gamma z_a/2$. The kernel $r(x; s)$ in the lossy case is computed in a similar method as in the lossless case. If $|\zeta_a| < \theta/2$, namely, the amplifier is located in the anomalous fiber segment, the resulting expression for kernel is

$$r(x; s) = \frac{1}{(2\pi)^2} \frac{Ge^{iC_0 x}}{(sx + 2iG\theta)(sx - 2iG(1 - \theta))}$$

$$\cdot \left[e^{G(4\zeta_a - 2\theta + 1)} sx \frac{\sin(sx - 2iG(1-\theta))}{\sinh(2G)} \right.$$

$$\left. + i\theta e^{i(4\zeta_a - 2\theta + 1)\frac{sx}{2\theta}} (sx - 2iG(1-\theta)) \right] \tag{8.22}$$

Note that in Eq.(8.22), unlike the lossless case, the kernel $r(x;s)$ is complex and is explicitly dependent on the parameters θ, Γ, z_a and ζ_a in a nontrivial way. However, one can still write

$$\lim_{s \to 0} r(x;s) = \frac{1}{(2\pi)^2} \tag{8.23}$$

and moreover

$$\lim_{G \to 0} r(x;s) = \frac{1}{(2\pi)^2} \frac{\sin(sx)}{sx} \tag{8.24}$$

which means that as $z_a \to 0$, Eq.(7.82) is recovered. The effect of nonlinearity on quasi-linear transmission with $g(\zeta) \neq 1$ is analyzed by the asymptotic expansion of the nonlinear terms in the GT equations for large s. The study will now be split into the following four cases, for $\theta = 1/2$, depending on the position of the amplifiers.

8.2.2.1 Case-I: $\zeta_a = 0$

This locates the amplifier in the middle of the anomalous GVD segments. In this case, the kernels $r(x;s)$ and $R(t_1, t_2; s)$ are

$$r(x;s) = \frac{1}{(2\pi)^2} \frac{G}{x^2 s^2 + G^2} \left[sx \frac{\sin sx}{\sinh G} + isx \left\{ 1 - \frac{\cos sx}{\cosh G} \right\} + G \right] \tag{8.25}$$

and

$$R(t_1, t_2; s) = \frac{G}{(2\pi)^2} \left[\left(\frac{I_{2S}}{\sinh G} + I_1 G \right) + i \left(I_2 - \frac{I_{2C}}{\cosh G} \right) \right] \tag{8.26}$$

where

$$I_1(t_1, t_2; s) = \int_{-\infty}^{\infty} \int_{-\infty}^{\infty} \frac{e^{i(\omega_1 t_1 + \omega_2 t_2)}}{G^2 + (s\omega_1\omega_2)^2} d\omega_1 d\omega_2 \tag{8.27}$$

$$I_2(t_1, t_2; s) = \int_{-\infty}^{\infty} \int_{-\infty}^{\infty} \frac{e^{i(\omega_1 t_1 + \omega_2 t_2)}}{G^2 + (s\omega_1\omega_2)^2} s\omega_1\omega_2 d\omega_1 d\omega_2 \tag{8.28}$$

$$I_{2S}(t_1, t_2; s) = \int_{-\infty}^{\infty} \int_{-\infty}^{\infty} \frac{e^{i(\omega_1 t_1 + \omega_2 t_2)}}{G^2 + (s\omega_1\omega_2)^2} s\omega_1\omega_2 \sin(s\omega_1\omega_2)\, d\omega_1 d\omega_2 \tag{8.29}$$

$$I_{2C}(t_1,t_2;s)=\int_{-\infty}^{\infty}\int_{-\infty}^{\infty}\frac{e^{i(\omega_1t_1+\omega_2t_2)}}{G^2+(s\omega_1\omega_2)^2}s\omega_1\omega_2\cos(s\omega_1\omega_2)\,d\omega_1d\omega_2 \quad (8.30)$$

The asymptotic expansion of the integrals in the kernel $R(t_1,t_2;s)$ for large s gives

$$R(t_1,t_2;s)\sim\frac{P_0}{s}(\log s-\log|t_1t_2|)-\frac{P_1}{s}-i\frac{P_2}{s}\text{sgn}(t_1t_2) \quad (8.31)$$

where $\text{sgn}(x)$ is the signum function defined as

$$\text{sgn}(t)=\frac{1}{i\pi}\int_{-\infty}^{\infty}\frac{e^{i\omega t}}{\omega}d\omega \quad (8.32)$$

and

$$P_0=\frac{G}{2\pi}\left(\frac{e^{-G}}{\sin G}+1\right) \quad (8.33)$$

$$P_1=\frac{G}{4\pi}\frac{1}{\sin G}\left[e^GI_{G_1}+(3\gamma+\log G-I_{G_2})e^{-G}\right]+\frac{G}{2\pi}(2\gamma+\log G) \quad (8.34)$$

$$P_2=\frac{G}{4} \quad (8.35)$$

with

$$I_{G_1}=\int_G^{\infty}\frac{e^{-x}}{x}dx \quad (8.36)$$

$$I_{G_2}=\int_0^G\frac{e^x-1}{x}dx \quad (8.37)$$

Note that

$$\lim_{G\to 0}P_0(G)=\frac{1}{2\pi} \quad (8.38)$$

$$\lim_{G\to 0}P_1(G)=\frac{\gamma}{2\pi} \quad (8.39)$$

$$\lim_{G\to 0}P_2(G)=0 \quad (8.40)$$

so that these reduce to the lossless case as Γ approaches zero. Thus, the GT equation, given by Eq.(8.1), for large s, reduces to

$$i\frac{\partial U}{\partial z}+\frac{\delta_a}{2}\frac{\partial^2U}{\partial t^2}+\frac{1}{s}[(P_0\log s-P_1)J_1(z,t)$$
$$-P_0J_2(z,t)-iP_2J_3(z,t)]=0 \quad (8.41)$$

where

$$J_3(z,t)=\int_{-\infty}^{\infty}\int_{-\infty}^{\infty}\text{sgn}(t_1t_2)U(z,t+t_1)U(z,t+t_2)$$
$$\cdot U^*(z,t+t_1+t_2)dt_1dt_2 \quad (8.42)$$

In the Fourier domain, (8.41) transforms to

$$i\frac{\partial \hat{U}}{\partial z} - \frac{\delta_a \omega^2}{2}\hat{U} + \frac{1}{s}[(P_0 \log s - P_1)\hat{J}_1(z,\omega)$$
$$-P_0 \hat{J}_2(z,\omega) - iP_2 \hat{J}_3(z,\omega)] = 0 \tag{8.43}$$

where

$$\hat{J}_3(z,\omega) = \int_{-\infty}^{\infty}\int_{-\infty}^{\infty} \frac{1}{\omega_1 \omega_2}\hat{U}(z,\omega+\omega_1)\hat{U}(z,\omega+\omega_2)$$
$$\cdot \hat{U}^*(z,\omega+\omega_1+\omega_2)d\omega_1 d\omega_2 \tag{8.44}$$

Note that in Eq.(8.44), the integral represents the Cauchy's principal value. Now, from Eq.(8.43) observe that

$$\frac{\partial}{\partial z}|\hat{U}|^2 = \frac{P_2}{s}H(z,\omega) \tag{8.45}$$

with

$$H(z,\omega) = \hat{J}_3(z,\omega)\hat{U}^*(z,\omega) + \hat{J}_3^*(z,\omega)\hat{U}(z,\omega) \tag{8.46}$$

For large s, and moderate z, one can write

$$\left|\hat{U}(z,\omega)\right|^2 \approx \left|\hat{U}(0,\omega)\right|^2 + \frac{P_2 z}{s}H(0,\omega) \tag{8.47}$$

and, thus, the total spectral intensity does not remain constant in this case.

8.2.2.2 Case-II: $\zeta_a = -1/2$

In this case, the amplifier is positioned in the middle of the normal GVD segment. The kernels of the GT equations, in this case, reduce to

$$r(x;s) = \frac{1}{(2\pi)^2}\frac{G}{x^2 s^2 + G^2}\left[sx\frac{\sin sx}{\sinh G} - isx\left\{1 - \frac{\cos sx}{\cosh G}\right\} + G\right] \tag{8.48}$$

and

$$R(t_1,t_2;s) = \frac{G}{(2\pi)^2}\left[\left(\frac{I_{2S}}{\sinh G} + I_1 G\right) - i\left(I_2 - \frac{I_{2C}}{\cosh G}\right)\right] \tag{8.49}$$

where the parameters are as the same as before. The only difference here is that imaginary part is negative. Thus, in the asymptotic state,

$$R(t_1,t_2;s) \sim \frac{P_0}{s}\left(\log s - \log|t_1 t_2|\right) - \frac{P_1}{s} + i\frac{P_2}{s}\mathrm{sgn}\,(t_1 t_2) \tag{8.50}$$

The intensity satisfies

$$\frac{\partial}{\partial z}|\hat{U}|^2 = -\frac{P_2}{s}H(z,\omega) \tag{8.51}$$

Finally, for $s \gg 1$ and moderate z,

$$\left|\hat{U}(z,\omega)\right|^2 \approx \left|\hat{U}(0,\omega)\right|^2 - \frac{P_2 z}{s}H(0,\omega) \tag{8.52}$$

So, the total spectral intensity does not stay conserved here, too.

8.2.2.3 Case-III: $\zeta_a = -1/4$

Here, the amplifier is placed at the boundary between the anomalous and normal GVD segment. In this case, the kernels are given by

$$\begin{aligned} r(x;s) = \frac{1}{(2\pi)^2}\frac{G}{x^2s^2+G^2}\Bigg[&\left\{\left(\frac{e^G}{\sinh G}-1\right)sx\sin sx + G\cos sx\right\} \\ &- i\left\{\left(\frac{e^{-G}}{\cosh G}-1\right)sx\cos sx + G\sin sx\right\}\Bigg] \end{aligned} \tag{8.53}$$

and

$$\begin{aligned} R(t_1,t_2;s) = \frac{G}{(2\pi)^2}\Bigg[&\left\{\left(\frac{e^{-G}}{\sinh G}-1\right)I_{2S} + GI_{1C}\right\} \\ &- i\left\{\left(\frac{e^{-G}}{\cosh G}-1\right)I_{2C} + GI_{1S}\right\}\Bigg] \end{aligned} \tag{8.54}$$

where

$$I_{1S}(t_1,t_2;s) = \int_{-\infty}^{\infty}\int_{-\infty}^{\infty}\frac{e^{i(\omega_1t_1+\omega_2t_2)}\sin s\omega_1\omega_2}{G^2+(s\omega_1\omega_2)^2}d\omega_1d\omega_2 \tag{8.55}$$

$$I_{1C}(t_1,t_2;s) = \int_{-\infty}^{\infty}\int_{-\infty}^{\infty}\frac{e^{i(\omega_1t_1+\omega_2t_2)}\cos s\omega_1\omega_2}{G^2+(s\omega_1\omega_2)^2}d\omega_1d\omega_2 \tag{8.56}$$

For large s,

$$R(t_1,t_2;s) \sim \frac{Q_0}{s}\left(\log s - \log|t_1t_2|\right) - \frac{Q_1}{s} \tag{8.57}$$

where

$$Q_0 = \frac{G}{2\pi}\frac{e^{-2G}}{\sinh G} \tag{8.58}$$

$$Q_1 = \frac{G}{4\pi}\frac{e^{-G}}{\sinh G}\left[e^G I_{G_1} + (3\gamma + \log G - I_{G_2})\,e^{-G}\right] - \frac{G}{2\pi}e^G I_{G_1} \tag{8.59}$$

Here, also

$$\lim_{G\to 0} Q_0(G) = \frac{1}{2\pi} \tag{8.60}$$

$$\lim_{G\to 0} Q_1(G) = \frac{\gamma}{2\pi} \tag{8.61}$$

so that, once again, as Γ approaches zero, it collapses to the lossless case. The GT equations, in this case, reduce to

$$i\frac{\partial U}{\partial z} + \frac{\delta_a}{2}\frac{\partial^2 U}{\partial t^2} + \frac{1}{s}\left[(Q_0 \log s - Q_1)J_1(z,t) - Q_0 J_2(z,t)\right] = 0 \tag{8.62}$$

while, in the Fourier domain,

$$i\frac{\partial \hat{U}}{\partial z} - \frac{\delta_a \omega^2}{2}\hat{U} + \frac{1}{s}\left[(Q_0 \log s - Q_1)\hat{J}_1(z,\omega) - Q_0 \hat{J}_2(z,\omega)\right] = 0 \tag{8.63}$$

In this case, note that the spectral intensity stays constant as from Eq.(8.63):

$$\frac{\partial}{\partial z}|\hat{U}|^2 = 0 \tag{8.64}$$

and also from Eq.(8.63), solution (8.19) is recovered where now

$$\begin{aligned} \psi\left[\left|\hat{U}(z,\omega)\right|^2\right] = \frac{1}{s}\Bigg[(Q_0 \log s - Q_1)\left|\hat{J}_1(z,\omega)\right|^2 \\ - \frac{Q_0}{\pi}\int_{-\infty}^{\infty}\left|\hat{J}_2(z,\omega')\right|^2 h(\omega-\omega')\,d\omega'\Bigg] \end{aligned} \tag{8.65}$$

8.2.2.4 Case-IV: $\zeta_a = 1/4$

Here, the amplifier is placed at the boundary between the normal and anomalous GVD segment. The kernels reduce to

$$\begin{aligned} r(x;s) = \frac{1}{(2\pi)^2}\frac{G}{x^2s^2+G^2}\Bigg[\left\{\left(\frac{e^G}{\sinh G}-1\right)sx\sin sx + G\cos sx\right\} \\ + i\left\{\left(\frac{e^{-G}}{\cosh G}-1\right)sx\cos sx + G\sin sx\right\}\Bigg] \end{aligned} \tag{8.66}$$

and

$$\begin{aligned} R(t_1,t_2;s) = \frac{G}{(2\pi)^2}\Bigg[\left\{\left(\frac{e^{-G}}{\sinh G}-1\right)I_{2S} + GI_{1C}\right\} \\ + i\left\{\left(\frac{e^{-G}}{\cosh G}-1\right)I_{2C} + GI_{1S}\right\}\Bigg] \end{aligned} \tag{8.67}$$

The only difference in this case from that of the previous one is the imaginary part of the kernel with an opposite sign. But, it was shown in the previous subsection that the imaginary part does not make any contribution to the dynamics of quasi-linear pulses, the sum of the spectral intensities is again conserved in this case during the pulse propagation. Figure 8.2 shows the plot of a quasi-linear Gaussian pulse in a lossy case, where $G = 0.5$ in the frequency domain.

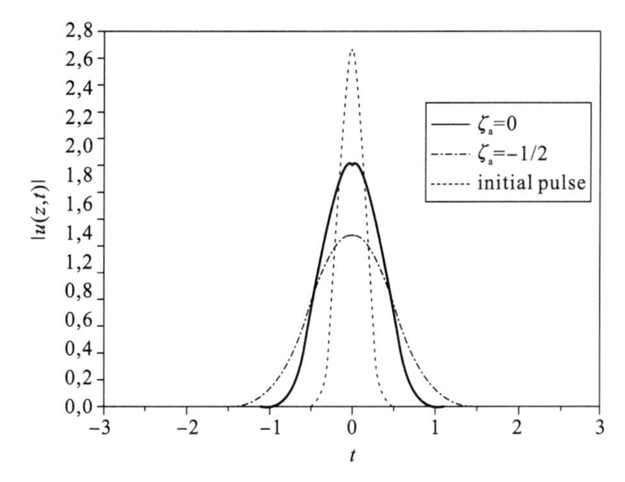

Fig. 8.2 Shape of the quasi-linear Gaussian pulse after propagation of $z = 20$ and the initial profile (dotted curve) for $s = 100$, $\langle d \rangle = 0$, and $G = 0.5$ in frequency domain.

8.3 Birefringent fibers

The corresponding GTE, in the Fourier domain, for the DM-VNLSE are

$$
\begin{aligned}
&i\frac{\partial \hat{U}}{\partial z} - \frac{\delta_a}{2}\omega^2 \hat{U} + \int_{-\infty}^{\infty}\int_{-\infty}^{\infty} r(\omega_1\omega_2)\hat{U}(z,\omega+\omega_2) \\
&\cdot \Big[\hat{U}(z,\omega+\omega_1)\hat{U}^*(z,\omega+\omega_1+\omega_2) \\
&+ \alpha\hat{V}(z,\omega+\omega_1)\hat{V}^*(z,\omega+\omega_1+\omega_2)\Big]\, d\omega_1 d\omega_2 = 0
\end{aligned} \tag{8.68}
$$

and

$$
\begin{aligned}
&i\frac{\partial \hat{V}}{\partial z} - \frac{\delta_a}{2}\omega^2 \hat{V} + \int_{-\infty}^{\infty}\int_{-\infty}^{\infty} r(\omega_1\omega_2)\hat{V}(z,\omega+\omega_2) \\
&\cdot \Big[\hat{V}(z,\omega+\omega_1)\hat{V}^*(z,\omega+\omega_1+\omega_2) \\
&+ \alpha\hat{U}(z,\omega+\omega_1)\hat{U}^*(z,\omega+\omega_1+\omega_2)\Big]\, d\omega_1 d\omega_2 = 0
\end{aligned} \tag{8.69}
$$

The GTE in the corresponding temporal domain are

$$i\frac{\partial U}{\partial z}+\frac{\delta_a}{2}\frac{\partial^2 U}{\partial t^2}+g(z)\int_{-\infty}^{\infty}\int_{-\infty}^{\infty}R(t_1,t_2)U(z,t_1+t_2)$$
$$\cdot\left[U(z,t+t_1)U^*(z,t+t_1+t_2)\right.$$
$$\left.+\,\alpha V(z,t+t_1)V^*(z,t+t_1+t_2)\right]dt_1dt_2=0 \quad (8.70)$$

and

$$i\frac{\partial V}{\partial z}+\frac{\delta_a}{2}\frac{\partial^2 V}{\partial t^2}+g(z)\int_{-\infty}^{\infty}\int_{-\infty}^{\infty}R(t_1,t_2)V(z,t_1+t_2)$$
$$\cdot\left[V(z,t+t_1)V^*(z,t+t_1+t_2)\right.$$
$$\left.+\,\alpha U(z,t+t_1)U^*(z,t+t_1+t_2)\right]dt_1dt_2=0 \quad (8.71)$$

Again, the study will be split into two subsections that deal with the lossless and the lossy cases, just as in the previous section.

8.3.1 Lossless system

Assuming that $\hat{U}(z,\omega)$ and $\hat{V}(z,\omega)$ depend weakly on s, the following asymptotic expansions of the nonlinear terms from the GTE equation is obtained:

$$i\frac{\partial U}{\partial z}+\frac{\delta_a}{2}\frac{\partial^2 U}{\partial t^2}+\frac{1}{2\pi s}\left[(\log s-\gamma)J_1^{(1)}(z,t)-J_2^{(1)}(z,t)\right]$$
$$+\,\frac{\alpha}{2\pi s}\left[(\log s-\gamma)K_1^{(1)}(z,t)-K_2^{(1)}(z,t)\right]=0 \quad (8.72)$$

and

$$i\frac{\partial V}{\partial z}+\frac{\delta_a}{2}\frac{\partial^2 V}{\partial t^2}+\frac{1}{2\pi s}\left[(\log s-\gamma)J_1^{(2)}(z,t)-J_2^{(2)}(z,t)\right]$$
$$+\,\frac{\alpha}{2\pi s}\left[(\log s-\gamma)K_1^{(2)}(z,t)-K_2^{(2)}(z,t)\right]=0 \quad (8.73)$$

where

$$J_1^{(1)}(z,t)=\int_{-\infty}^{\infty}\int_{-\infty}^{\infty}U(z,t+t_1)\,U(z,t+t_2)$$
$$\cdot U^*(z,t+t_1+t_2)\,dt_1dt_2 \quad (8.74)$$

$$J_2^{(1)}(z,t)=\int_{-\infty}^{\infty}\int_{-\infty}^{\infty}\log|t_1t_2|\,U(z,t+t_1)U(z,t+t_2)$$
$$\cdot U^*(z,t+t_1+t_2)dt_1dt_2 \quad (8.75)$$

$$K_1^{(1)}(z,t)=\int_{-\infty}^{\infty}\int_{-\infty}^{\infty}V(z,t+t_2)U(z,t+t_1)V^*(z,t+t_1+t_2)dt_1dt_2 \quad (8.76)$$

$$K_2^{(1)}(z,t) = \int_{-\infty}^{\infty}\int_{-\infty}^{\infty} \log|t_1 t_2| V(z,t+t_2)U(z,t+t_1) \cdot V^*(z,t+t_1+t_2)dt_1 dt_2 \tag{8.77}$$

$$J_1^{(2)}(z,t) = \int_{-\infty}^{\infty}\int_{-\infty}^{\infty} V(z,t+t_1)V(z,t+t_2)V^*(z,t+t_1+t_2)dt_1 dt_2 \tag{8.78}$$

$$J_2^{(2)}(z,t) = \int_{-\infty}^{\infty}\int_{-\infty}^{\infty} \log|t_1 t_2| V(z,t+t_1)V(z,t+t_2) \cdot V^*(z,t+t_1+t_2)dt_1 dt_2 \tag{8.79}$$

$$K_1^{(2)}(z,t) = \int_{-\infty}^{\infty}\int_{-\infty}^{\infty} V(z,t+t_2)U(z,t+t_1)U^*(z,t+t_1+t_2)dt_1 dt_2 \tag{8.80}$$

$$K_2^{(2)}(z,t) = \int_{-\infty}^{\infty}\int_{-\infty}^{\infty} \log|t_1 t_2| V(z,t+t_2)U(z,t+t_1) \cdot U^*(z,t+t_1+t_2)dt_1 dt_2 \tag{8.81}$$

In the Fourier domain, (8.72) and (8.73) respectively transform to

$$i\frac{\partial \hat{U}}{\partial z} - \frac{\delta_a \omega^2}{2}\hat{U} + \frac{1}{2\pi s}\left[(\log s - \gamma)\hat{J}_1^{(1)}(z,\omega) - \hat{J}_2^{(1)}(z,\omega)\right] + \frac{\alpha}{2\pi s}\left[(\log s - \gamma)\hat{K}_1^{(1)}(z,\omega) - \hat{K}_2^{(1)}(z,\omega)\right] = 0 \tag{8.82}$$

and

$$i\frac{\partial \hat{V}}{\partial z} - \frac{\delta_a \omega^2}{2}\hat{V} + \frac{1}{2\pi s}\left[(\log s - \gamma)\hat{J}_1^{(2)}(z,\omega) - \hat{J}_2^{(2)}(z,\omega)\right] + \frac{\alpha}{2\pi s}\left[(\log s - \gamma)\hat{K}_1^{(2)}(z,\omega) - \hat{K}_2^{(2)}(z,\omega)\right] = 0 \tag{8.83}$$

where

$$\hat{J}_1^{(1)}(z,\omega) = \left|\hat{U}(z,\omega)\right|^2 \hat{U}(z,\omega) \tag{8.84}$$

$$\hat{J}_2^{(1)}(z,\omega) = \frac{1}{\pi}\hat{U}(z,\omega)\int_{-\infty}^{\infty}\left|\hat{U}(z,\omega')\right|^2 h(\omega'-\omega)\,d\omega' \tag{8.85}$$

$$\hat{K}_1^{(1)}(z,\omega) = \left|\hat{V}(z,\omega)\right|^2 \hat{U}(z,\omega) \tag{8.86}$$

$$\hat{K}_2^{(1)}(z,\omega) = \frac{1}{\pi}\hat{U}(z,\omega)\int_{-\infty}^{\infty}\left|\hat{V}(z,\omega')\right|^2 h(\omega'-\omega)\,d\omega' \tag{8.87}$$

$$\hat{J}_1^{(2)}(z,\omega) = \left|\hat{V}(z,\omega)\right|^2 \hat{V}(z,\omega) \tag{8.88}$$

$$\hat{J}_2^{(2)}(z,\omega) = \frac{1}{\pi}\hat{V}(z,\omega)\int_{-\infty}^{\infty}\left|\hat{V}(z,\omega')\right|^2 h(\omega'-\omega)\,d\omega' \tag{8.89}$$

$$\hat{K}_1^{(2)}(z,\omega) = \left|\hat{U}(z,\omega)\right|^2 \hat{V}(z,\omega) \tag{8.90}$$

$$\hat{K}_2^{(2)}(z,\omega) = \frac{1}{\pi}\hat{V}(z,\omega)\int_{-\infty}^{\infty}\left|\hat{U}(z,\omega')\right|^2 h\left(\omega'-\omega\right)d\omega' \tag{8.91}$$

One can see from Eqs.(8.82) and (8.83) that the total spectral intensity given by $|\hat{U}(z,\omega)|^2 + |\hat{V}(z,\omega)|^2$ is preserved during the pulse propagation, namely,

$$\frac{\partial}{\partial z}\left(|\hat{U}|^2 + |\hat{V}|^2\right) = 0 \tag{8.92}$$

Also, the solutions of Eqs.(8.82) and (8.83) are

$$\hat{U}(z,\omega) = \hat{U}(z,0)\exp\left\{-i\frac{\delta_a\omega^2}{2}z + i\psi\left[|\hat{U}(0,\omega)|^2\right]z + i\alpha\psi\left[|\hat{V}(0,\omega)|^2\right]z\right\} \tag{8.93}$$

and

$$\hat{V}(z,\omega) = \hat{V}(z,0)\exp\left\{-i\frac{\delta_a\omega^2}{2}z + i\psi\left[|\hat{V}(0,\omega)|^2\right]z + i\alpha\psi\left[|\hat{U}(0,\omega)|^2\right]z\right\} \tag{8.94}$$

respectively, where

$$\psi\left[|\hat{U}(z,\omega)|^2\right] = \frac{1}{2\pi s}\left[(\log s - \gamma)|\hat{U}(z,\omega)|^2 - \frac{1}{\pi}\int_{-\infty}^{\infty}|\hat{U}(z,\omega)|^2 h(\omega'-\omega)d\omega'\right] \tag{8.95}$$

with a similar expression for $\psi[|\hat{V}(z,\omega)|^2]$.

8.3.2 lossy system

For lossy case, the study will be further split into the following four cases, for $\theta = 1/2$, depending on the positions of the amplifiers.

8.3.2.1 Case-I: $\zeta_a = 0$

This locates the amplifier in the middle of the anomalous GVD segments. In this case, the kernels $r(x;s)$ and $R(t_1,t_2;s)$ are as in Eqs.(8.25) and (8.26), respectively. The GT equations, given by Eqs.(8.72) and (8.73), for large s, reduces to [8]

$$\begin{aligned} & i\frac{\partial U}{\partial z} + \frac{\delta_a}{2}\frac{\partial^2 U}{\partial t^2} + \frac{1}{s}[(P_0 \log s - P_1)J_1^{(1)}(z,t) \\ & -P_0 J_2^{(1)}(z,t) - iP_2 J_3^{(1)}(z,t)] + \frac{\alpha}{s}[(P_0 \log s - P_1)K_1^{(1)}(z,t) \\ & -P_0 K_2^{(1)}(z,t) - iP_2 K_3^{(1)}(z,t)] = 0 \end{aligned} \tag{8.96}$$

and

$$\begin{aligned} & i\frac{\partial V}{\partial z} + \frac{\delta_a}{2}\frac{\partial^2 V}{\partial t^2} + \frac{1}{s}[(P_0 \log s - P_1)J_1^{(2)}(z,t) \\ & -P_0 J_2^{(2)}(z,t) - iP_2 J_3^{(2)}(z,t)] + \frac{\alpha}{s}[(P_0 \log s - P_1)K_1^{(2)}(z,t) \\ & -P_0 K_2^{(2)}(z,t) - iP_2 K_3^{(2)}(z,t)] = 0 \end{aligned} \tag{8.97}$$

where

$$\begin{aligned} J_3^{(1)}(z,t) = & \int_{-\infty}^{\infty}\int_{-\infty}^{\infty} \operatorname{sgn}(t_1 t_2) U(z,t+t_1) U(z,t+t_2) \\ & \cdot U^*(z,t+t_1+t_2) dt_1 dt_2 \end{aligned} \tag{8.98}$$

$$\begin{aligned} J_3^{(2)}(z,t) = & \int_{-\infty}^{\infty}\int_{-\infty}^{\infty} \operatorname{sgn}(t_1 t_2) V(z,t+t_1) V(z,t+t_2) \\ & \cdot V^*(z,t+t_1+t_2) dt_1 dt_2 \end{aligned} \tag{8.99}$$

$$\begin{aligned} K_3^{(1)}(z,t) = & \int_{-\infty}^{\infty}\int_{-\infty}^{\infty} \operatorname{sgn}(t_1 t_2) V(z,t+t_1) U(z,t+t_2) \\ & \cdot V^*(z,t+t_1+t_2) dt_1 dt_2 \end{aligned} \tag{8.100}$$

$$\begin{aligned} K_3^{(2)}(z,t) = & \int_{-\infty}^{\infty}\int_{-\infty}^{\infty} \operatorname{sgn}(t_1 t_2) U(z,t+t_1) V(z,t+t_2) \\ & \cdot U^*(z,t+t_1+t_2) dt_1 dt_2 \end{aligned} \tag{8.101}$$

In the Fourier domain, Eqs.(8.96) and (8.97) respectively reduce to [8]

$$\begin{aligned} & i\frac{\partial \hat{U}}{\partial z} - \frac{\delta_a \omega^2}{2}\hat{U} + \frac{1}{s}[(P_0 \log s - P_1)\hat{J}_1^{(1)}(z,\omega) - P_0 \hat{J}_2^{(1)}(z,\omega) \\ & -iP_2 \hat{J}_3^{(1)}(z,\omega)] + \frac{\alpha}{s}[(P_0 \log s - P_1)\hat{K}_1^{(1)}(z,\omega) \\ & -P_0 \hat{K}_2^{(1)}(z,\omega) - iP_2 \hat{K}_3^{(1)}(z,\omega)] = 0 \end{aligned} \tag{8.102}$$

and

$$i\frac{\partial \hat{V}}{\partial z} - \frac{\delta_a \omega^2}{2}\hat{V} + \frac{1}{s}[(P_0 \log s - P_1)\hat{J}_1^{(2)}(z,\omega) - P_0 \hat{J}_2^{(2)}(z,\omega)$$

$$-iP_2\hat{J}_3^{(2)}(z,\omega)] + \frac{\alpha}{s}[(P_0 \log s - P_1)\hat{K}_1^{(2)}(z,\omega)$$
$$-P_0\hat{K}_2^{(2)}(z,\omega) - iP_2\hat{K}_3^{(2)}(z,\omega)] = 0 \tag{8.103}$$

where

$$\hat{J}_3^{(1)}(z,\omega) = \int_{-\infty}^{\infty}\int_{-\infty}^{\infty} \frac{1}{\omega_1\omega_2}\hat{U}(z,\omega+\omega_1)\hat{U}(z,\omega+\omega_2)$$
$$\cdot\hat{U}^*(z,\omega+\omega_1+\omega_2)d\omega_1 d\omega_2 \tag{8.104}$$

$$\hat{J}_3^{(2)}(z,\omega) = \int_{-\infty}^{\infty}\int_{-\infty}^{\infty} \frac{1}{\omega_1\omega_2}\hat{V}(z,\omega+\omega_1)\hat{V}(z,\omega+\omega_2)$$
$$\cdot\hat{V}^*(z,\omega+\omega_1+\omega_2)d\omega_1 d\omega_2 \tag{8.105}$$

$$\hat{K}_3^{(1)}(z,\omega) = \int_{-\infty}^{\infty}\int_{-\infty}^{\infty} \frac{1}{\omega_1\omega_2}\hat{V}(z,\omega+\omega_1)\hat{U}(z,\omega+\omega_2)$$
$$\cdot\hat{V}^*(z,\omega+\omega_1+\omega_2)d\omega_1 d\omega_2 \tag{8.106}$$

$$\hat{K}_3^{(2)}(z,\omega) = \int_{-\infty}^{\infty}\int_{-\infty}^{\infty} \frac{1}{\omega_1\omega_2}\hat{U}(z,\omega+\omega_1)\hat{V}(z,\omega+\omega_2)$$
$$\cdot\hat{U}^*(z,\omega+\omega_1+\omega_2)d\omega_1 d\omega_2 \tag{8.107}$$

In Eqs.(8.104)–(8.107), the integrals represent the Cauchy's principal values. Now, from Eqs.(8.102) and (8.103) observe that

$$\frac{\partial}{\partial z}\left(|\hat{U}|^2 + |\hat{V}|^2\right) = \frac{P_2}{s}\left\{H^{(1)}(z,\omega) + \alpha H^{(2)}(z,\omega)\right\} \tag{8.108}$$

with

$$H^{(1)}(z,\omega) = \hat{J}_3^{(1)}(z,\omega)\hat{U}^*(z,\omega) + \hat{J}_3^{(1)*}(z,\omega)\hat{U}(z,\omega) \tag{8.109}$$

and

$$H^{(2)}(z,\omega) = \hat{K}_3^{(1)}(z,\omega)\hat{V}^*(z,\omega) + \hat{K}_3^{(1)*}(z,\omega)\hat{V}(z,\omega) \tag{8.110}$$

For large s, and moderate z, one can write

$$|\hat{U}(z,\omega)|^2 + |\hat{V}(z,\omega)|^2$$
$$\approx |\hat{U}(0,\omega)|^2 + |\hat{V}(0,\omega)|^2 + \frac{P_2 z}{s}\{H^{(1)}(0,\omega) + \alpha H^{(1)}(0,\omega)\} \tag{8.111}$$

and, thus, the total spectral intensity does not remain constant in this case.

8.3.2.2 Case-II: $\zeta_a = -1/2$

In this case, the amplifier is positioned in the middle of the normal GVD segment. Here, the kernels are respectively given by Eqs. (8.48) and (8.49). In this case also,

$$\frac{\partial}{\partial z}\left(|\hat{U}|^2 + |\hat{V}|^2\right) = -\frac{P_2}{s}\{H^{(1)}(z,\omega) + \alpha H^{(2)}(z,\omega)\} \tag{8.112}$$

Finally, for $s \gg 1$ and moderate z leads to

$$|\hat{U}(z,\omega)|^2 + |\hat{V}(z,\omega)|^2$$
$$\approx |\hat{U}(0,\omega)|^2 + |\hat{V}(0,\omega)|^2 - \frac{P_2 z}{s}\{H^{(1)}(0,\omega) + \alpha H^{(1)}(0,\omega)\} \tag{8.113}$$

So, the total spectral intensity does not stay conserved here, too.

8.3.2.3 Case-III: $\zeta_a = -1/4$

Here, the amplifier is placed at the boundary between the anomalous and normal GVD segment and the kernels are the same as in Eqs.(8.53) and (8.54). In this case, the GTE given by Eqs.(8.96) and (8.97) reduces to

$$i\frac{\partial U}{\partial z} + \frac{\delta_a}{2}\frac{\partial^2 U}{\partial t^2} + \frac{1}{s}\left[(Q_0 \log s - Q_1)J_1^{(1)}(z,t) - Q_0 J_2^{(1)}(z,t)\right]$$
$$+\frac{\alpha}{s}\left[(Q_0 \log s - Q_1)K_1^{(1)}(z,t) - Q_0 K_2^{(1)}(z,t)\right] = 0 \tag{8.114}$$

and

$$i\frac{\partial V}{\partial z} + \frac{\delta_a}{2}\frac{\partial^2 V}{\partial t^2} + \frac{1}{s}\left[(Q_0 \log s - Q_1)J_1^{(2)}(z,t) - Q_0 J_2^{(2)}(z,t)\right]$$
$$+\frac{\alpha}{s}\left[(Q_0 \log s - Q_1)K_1^{(2)}(z,t) - Q_0 K_2^{(2)}(z,t)\right] = 0 \tag{8.115}$$

while, in the Fourier domain,

$$i\frac{\partial \hat{U}}{\partial z} - \frac{\delta_a \omega^2}{2}\hat{U} + \frac{1}{s}\left[(Q_0 \log s - Q_1)\hat{J}_1^{(1)}(z,\omega) - Q_0 \hat{J}_2^{(1)}(z,\omega)\right]$$
$$+\frac{\alpha}{s}\left[(Q_0 \log s - Q_1)\hat{K}_1^{(1)}(z,\omega) - Q_0 \hat{K}_2^{(1)}(z,\omega)\right] = 0 \tag{8.116}$$

and

$$i\frac{\partial \hat{V}}{\partial z} - \frac{\delta_a \omega^2}{2}\hat{V} + \frac{1}{s}\left[(Q_0 \log s - Q_1)\hat{J}_1^{(2)}(z,\omega) - Q_0 \hat{J}_2^{(2)}(z,\omega)\right]$$

$$+\frac{\alpha}{s}\left[(Q_0\log s - Q_1)\hat{K}_1^{(2)}(z,\omega) - Q_0\hat{K}_2^{(2)}(z,\omega)\right] = 0 \tag{8.117}$$

In this case, that the total spectral intensity stays constant as from Eqs.(8.116) and (8.117):

$$\frac{\partial}{\partial z}\left(|\hat{U}|^2 + |\hat{V}|^2\right) = 0 \tag{8.118}$$

and also from Eqs.(8.116) and (8.117), solutions (8.93) and (8.94) respectively are recovered where now

$$\begin{aligned}\psi\left[|\hat{U}(z,\omega)|^2\right] = \frac{1}{s}\Bigg[&(Q_0\log s - Q_1)|\hat{J}_1^{(1)}(z,\omega)|^2 \\ &- \frac{Q_0}{\pi}\int_{-\infty}^{\infty}|\hat{J}_2^{(1)}(z,\omega')|^2 h(\omega-\omega')\,d\omega'\Bigg] \\ &+ \frac{\alpha}{s}\Bigg[(Q_0\log s - Q_1)|\hat{K}_1^{(1)}(z,\omega)|^2 \\ &- \frac{Q_0}{\pi}\int_{-\infty}^{\infty}|\hat{K}_2^{(1)}(z,\omega')|^2 h(\omega-\omega')\,d\omega'\Bigg]\end{aligned} \tag{8.119}$$

and

$$\begin{aligned}\psi\left[|\hat{V}(z,\omega)|^2\right] = \frac{1}{s}\Bigg[&(Q_0\log s - Q_1)|\hat{J}_1^{(2)}(z,\omega)|^2 \\ &- \frac{Q_0}{\pi}\int_{-\infty}^{\infty}|\hat{J}_2^{(2)}(z,\omega')|^2 h(\omega-\omega')\,d\omega'\Bigg] \\ &+ \frac{\alpha}{s}\Bigg[(Q_0\log s - Q_1)|\hat{K}_1^{(2)}(z,\omega)|^2 \\ &- \frac{Q_0}{\pi}\int_{-\infty}^{\infty}|\hat{K}_2^{(2)}(z,\omega')|^2 h(\omega-\omega')\,d\omega'\Bigg]\end{aligned} \tag{8.120}$$

8.3.2.4 Case-IV: $\zeta_a = 1/4$

Here, the amplifier is placed at the boundary between the normal and anomalous GVD segments. In this case, the kernels are given by Eqs.(8.66) and (8.67). The only difference in this case from that of the previous one is the imaginary part of the kernels with opposite signs. But, it was shown in the previous subsection that the imaginary part does not make any contribution to the dynamics of quasi-linear pulses, the sum of the spectral intensities is again conserved in this case during the pulse propagation.

8.4 Multiple channels

Recall that for DWDM systems, the GTE is given by

$$iu_{l,z} + \frac{D(z)}{2}u_{l,tt} + g(z)\left\{ |u_l|^2 + \sum_{m\neq l}^{N} \alpha_{lm} |u_m|^2 \right\}u_l = 0 \tag{8.121}$$

where $1 \leq l \leq N$ and models for bit-parallel WDM soliton transmission. Also, α_{lm} are known as the XPM coefficients. The corresponding GTE for the case of multiple channels in the Fourier domain is given by

$$\begin{aligned} i\frac{\partial \hat{U}_l}{\partial z} - \frac{\delta_a}{2}\omega^2\hat{U}_l + \int_{-\infty}^{\infty}\int_{-\infty}^{\infty} r(\omega_1\omega_2)\hat{U}_l(z,\omega_1+\omega_2) \\ \cdot\left[\hat{U}_l(z,\omega+\omega_1)\hat{U}_l^{\,*}(z,\omega+\omega_1+\omega_2)\right. \\ \left. + \sum_{l\neq m}^{N} \alpha_{lm}\hat{U}_m(z,\omega+\omega_1)\hat{U}_m^{\,*}(z,\omega+\omega_1+\omega_2)\right] d\omega_1 d\omega_2 = 0 \end{aligned} \tag{8.122}$$

while in the time domain, the GT equation is

$$\begin{aligned} i\frac{\partial U_l}{\partial z} + \frac{\delta_a}{2}\frac{\partial^2 U_l}{\partial t^2} + g(z)\int_{-\infty}^{\infty}\int_{-\infty}^{\infty} R(t_1,t_2)U_l(z,t_1+t_2) \\ \cdot\left[U_l(z,t+t_1)U_l^*(z,t+t_1+t_2)\right. \\ \left. + \sum_{l\neq m}^{N} \alpha_{lm}U_m(z,t+t_1)U_m^*(z,t+t_1+t_2)\right] dt_1 dt_2 = 0 \end{aligned} \tag{8.123}$$

where $1 \leq l \leq N$. Once again, as before the study will be split into two subsections namely the lossless and the lossy cases.

8.4.1 Lossless system

In a lossless system, namely $g(z) = 1$, the kernels are given by Eqs. (8.5) and (8.6). The following asymptotic expansion of the nonlinear term from the GTE is obtained:

$$\begin{aligned} i\frac{\partial U_l}{\partial z} + \frac{\delta_a}{2}\frac{\partial^2 U_l}{\partial t^2} + \frac{1}{2\pi s}\left[(\log s - \gamma)J_1^{(l)}(z,t) - J_2^{(l)}(z,t)\right] \\ + \frac{1}{2\pi s}\sum_{m\neq l}\alpha_{lm}\left[(\log s - \gamma)K_1^{(m)}(z,t) - K_2^{(m)}(z,t)\right] = 0 \end{aligned} \tag{8.124}$$

where

$$J_1^{(l)}(z,t) = \int_{-\infty}^{\infty}\int_{-\infty}^{\infty} U_l(z,t+t_1)U_l(z,t+t_2)U_l^*(z,t+t_1+t_2)dt_1dt_2 \quad (8.125)$$

$$J_2^{(l)}(z,t) = \int_{-\infty}^{\infty}\int_{-\infty}^{\infty} \log|t_1t_2|\, U_l(z,t+t_1)U_l(z,t+t_2) \cdot U_l^*(z,t+t_1+t_2)dt_1dt_2 \quad (8.126)$$

$$K_1^{(l)}(z,t) = \sum_{m\neq l}\alpha_{lm}\int_{-\infty}^{\infty}\int_{-\infty}^{\infty} U_l(z,t+t_2)U_m(z,t+t_1) \cdot U_m^*(z,t+t_1+t_2)dt_1dt_2 \quad (8.127)$$

$$K_2^{(l)}(z,t) = \sum_{m\neq l}\alpha_{lm}\int_{-\infty}^{\infty}\int_{-\infty}^{\infty} \log|t_1t_2|U_l(z,t+t_2)U_m(z,t+t_1) \cdot U_m^*(z,t+t_1+t_2)dt_1dt_2 \quad (8.128)$$

In the Fourier domain,

$$i\frac{\partial\hat{U}_l}{\partial z} - \frac{\delta_a\omega^2}{2}\hat{U}_l + \frac{1}{2\pi s}[(\log s-\gamma)\hat{J}_1^{(l)}(z,\omega) - \hat{J}_2^{(l)}(z,\omega)] + \frac{\alpha}{2\pi s}\sum_{m\neq l}\alpha_{lm}[(\log s-\gamma)\hat{K}_1^{(m)}(z,\omega) - \hat{K}_2^{(m)}(z,\omega)] = 0 \quad (8.129)$$

where

$$\hat{J}_1^{(l)}(z,\omega) = |\hat{U}_l(z,\omega)|^2\hat{U}_l(z,\omega) \quad (8.130)$$

$$\hat{J}_2^{(l)}(z,\omega) = \frac{1}{\pi}\hat{U}_l(z,\omega)\int_{-\infty}^{\infty}|\hat{U}_l(z,\omega')|^2h(\omega'-\omega)d\omega' \quad (8.131)$$

$$\hat{K}_1^{(l)}(z,\omega) = \hat{U}_l(z,\omega)\sum_{m\neq l}\alpha_{lm}|\hat{U}_m(z,\omega)|^2 \quad (8.132)$$

$$\hat{K}_2^{(l)}(z,\omega) = \frac{1}{\pi}\hat{U}_l(z,\omega)\sum_{m\neq l}\alpha_{lm}\int_{-\infty}^{\infty}|\hat{U}_m(z,\omega')|^2h(\omega'-\omega)d\omega' \quad (8.133)$$

Now, one can see from Eq.(8.129) that the total spectral intensity given by $\sum_{l=1}^{N}|\hat{U}_l(z,\omega)|^2$ is preserved during the pulse propagation. So,

$$\sum_{l=1}^{N}\frac{\partial}{\partial z}|\hat{U}_l|^2 = 0 \quad (8.134)$$

Thus, the solution of Eq.(8.129) is

$$\hat{U}_l(z,\omega) = \hat{U}_l(z,0)\exp\left\{ -i\frac{\delta_a\omega^2}{2}z + i\psi\left[|\hat{U}_l(0,\omega)|^2\right]z\right.$$

$$+ i \sum_{m \neq l} \alpha_{lm} \psi \left[|\hat{U}_m(0,\omega)|^2 \right] z \Big\} \tag{8.135}$$

where

$$\psi \left[|\hat{U}_l(z,\omega)|^2 \right] = \frac{1}{2\pi s} \left[(\log s - \gamma) |\hat{U}_l(z,\omega)|^2 \right.$$
$$\left. - \frac{1}{\pi} \int_{-\infty}^{\infty} |\hat{U}_l(z,\omega)|^2 h(\omega' - \omega) d\omega' \right] \tag{8.136}$$

8.4.2 Lossy system

The study in this case, will be similarly split into the following four cases depending on the position of the amplifiers.

8.4.2.1 Case-I: $\zeta_a = 0$

This locates the amplifier in the middle of the anomalous GVD segments. In this case, the kernels $r(x;s)$ and $R(t_1,t_2;s)$ as in Eqs.(8.25) and (8.26). The GT equations for large s reduce to

$$i\frac{\partial U_l}{\partial z} + \frac{\delta_a}{2}\frac{\partial^2 U_l}{\partial t^2} + \frac{1}{s}[(P_0 \log s - P_1) J_1^{(l)}(z,t) - P_0 J_2^{(l)}(z,t)$$
$$-iP_2 J_3^{(l)}(z,t)] + \frac{1}{s}\sum_{m \neq l} \alpha_{lm}[(P_0 \log s - P_1) K_1^{(m)}(z,t)$$
$$-P_0 K_2^{(m)}(z,t) - iP_2 K_3^{(m)}(z,t)] = 0 \tag{8.137}$$

where

$$J_3^{(l)}(z,t) = \int_{-\infty}^{\infty}\int_{-\infty}^{\infty} \operatorname{sgn}(t_1 t_2) U_l(z,t+t_1) U_l(z,t+t_2)$$
$$\cdot U_l^*(z,t+t_1+t_2) dt_1 dt_2 \tag{8.138}$$

$$K_3^{(m)}(z,t) = \sum_{m \neq l} \alpha_{lm} \int_{-\infty}^{\infty}\int_{-\infty}^{\infty} \operatorname{sgn}(t_1 t_2) U_l(z,t+t_2)$$
$$\cdot U_m(z,t+t_1) U_m^*(z,t+t_1+t_2) dt_1 dt_2 \tag{8.139}$$

In the Fourier domain, these equations are

$$i\frac{\partial \hat{U}_l}{\partial z} - \frac{\delta_a \omega^2}{2}\hat{U}_l + \frac{1}{s}[(P_0 \log s - P_1)\hat{J}_1^{(l)}(z,\omega) - P_0 \hat{J}_2^{(l)}(z,\omega)$$

$$-iP_2\hat{J}_3^{(l)}(z,\omega)] + \frac{1}{s}\sum_{m\neq l}\alpha_{lm}[(P_0\log s - P_1)\hat{K}_1^{(m)}(z,\omega)$$

$$-P_0\hat{K}_2^{(m)}(z,\omega) - iP_2\hat{K}_3^{(m)}(z,\omega)] = 0 \tag{8.140}$$

where

$$\hat{J}_3^{(l)}(z,\omega) = \int_{-\infty}^{\infty}\int_{-\infty}^{\infty}\frac{1}{\omega_1\omega_2}\hat{U}_l(z,\omega+\omega_1)\hat{U}_l(z,\omega+\omega_2)$$

$$\cdot\hat{U}_l^*(z,\omega+\omega_1+\omega_2)d\omega_1 d\omega_2 \tag{8.141}$$

$$\hat{K}_3^{(m)}(z,\omega) = \sum_{m\neq l}\alpha_{lm}\int_{-\infty}^{\infty}\int_{-\infty}^{\infty}\frac{1}{\omega_1\omega_2}\hat{U}_l(z,\omega+\omega_1)$$

$$\cdot\hat{U}_m(z,\omega+\omega_2)\hat{U}_m^*(z,\omega+\omega_1+\omega_2)d\omega_1 d\omega_2 \tag{8.142}$$

The integrals in Eqs.(8.141) and (8.142) represent the Cauchy's principal value. Now, from Eq.(8.140),

$$\sum_{l=1}^{N}\frac{\partial}{\partial z}\left(|\hat{U}_l|^2\right) = \frac{P_2}{s}\left\{H^{(l)}(z,\omega) + \sum_{m\neq l}\alpha_{lm}H^{(m)}(z,\omega)\right\} \tag{8.143}$$

with

$$H^{(l)}(z,\omega) = \hat{J}_3^{(l)}(z,\omega)\hat{U}_l^*(z,\omega) + \hat{J}_3^{(l)*}(z,\omega)\hat{U}_l(z,\omega) \tag{8.144}$$

and

$$H^{(m)}(z,\omega) = \hat{K}_3^{(m)}(z,\omega)\hat{U}_m^*(z,\omega) + \hat{K}_3^{(m)*}(z,\omega)\hat{U}_m(z,\omega) \tag{8.145}$$

For large s, one can write for moderate z,

$$\sum_{l=1}^{N}|\hat{U}_l(z,\omega)|^2$$

$$\approx \sum_{l=1}^{N}|\hat{U}(0,\omega)|^2 + \frac{P_2 z}{s}\left\{H^{(l)}(0,\omega) + \sum_{m\neq l}\alpha_{lm}H^{(1)}(0,\omega)\right\} \tag{8.146}$$

and, thus, the total spectral intensity does not remain constant in this case.

8.4.2.2 Case-II: $\zeta_a = -1/2$

In this case, the amplifier is positioned in the middle of the normal GVD segment. The kernels of the GTE in this case are given by Eqs.(8.48) and (8.49). Then, the sum of the intensities satisfy

$$\sum_{l=1}^{N} \frac{\partial}{\partial z} |\hat{U}_l|^2 = -\frac{P_2}{s}\left\{ H^{(l)}(z,\omega) + \sum_{m\neq l} \alpha_{lm} H^{(m)}(z,\omega) \right\} \tag{8.147}$$

Finally, for $s \gg 1$ and moderate z, one obtains

$$\sum_{l=1}^{N} |\hat{U}_l(z,\omega)|^2$$

$$\approx \sum_{l=1}^{N} |\hat{U}_l(0,\omega)|^2 - \frac{P_2 z}{s}\left\{ H^{(l)}(0,\omega) + \sum_{m\neq l} \alpha_{lm} H^{(1)}(0,\omega) \right\} \tag{8.148}$$

So, the total spectral intensity does not stay conserved here, too.

8.4.2.3 Case-III: $\zeta_a = -1/4$

Here, the amplifier is placed at the boundary between the anomalous and normal GVD segments. In this case, the kernels are given by Eqs.(8.53) and (8.54). Thus, the GTE, in this case, is reduced to

$$\begin{aligned} i\frac{\partial U_l}{\partial z} &+ \frac{\delta_a}{2}\frac{\partial^2 U_l}{\partial t^2} + \frac{1}{s}[(Q_0 \log s - Q_1) J_1^{(l)}(z,t) - Q_0 J_2^{(l)}(z,t)] \\ &+ \frac{1}{s}\sum_{m\neq l} \alpha_{lm}[(Q_0 \log s - Q_1) K_1^{(m)}(z,t) - Q_0 K_2^{(m)}(z,t)] = 0 \end{aligned} \tag{8.149}$$

so that, in the Fourier domain,

$$\begin{aligned} i\frac{\partial \hat{U}_l}{\partial z} &- \frac{\delta_a \omega^2}{2}\hat{U}_l + \frac{1}{s}[(Q_0 \log s - Q_1) \hat{J}_1^{(l)}(z,\omega) - Q_0 \hat{J}_2^{(l)}(z,\omega)] \\ &+ \frac{1}{s}\sum_{m\neq l} \alpha_{lm}[(Q_0 \log s - Q_1) \hat{K}_1^{(m)}(z,\omega) - Q_0 \hat{K}_2^{(m)}(z,\omega)] = 0 \end{aligned} \tag{8.150}$$

In this case, the total spectral intensity from all the channels stays constant since

$$\sum_{l=1}^{N} \frac{\partial}{\partial z} |\hat{U}_l|^2 = 0 \tag{8.151}$$

so that, from Eq.(8.150), it is possible to get

$$\begin{aligned} \psi\left[|\hat{U}_l(z,\omega)|^2 \right] = \frac{1}{s}\Bigg[&(Q_0 \log s - Q_1)|\hat{J}_1^{(l)}(z,\omega)|^2 \\ &- \frac{Q_0}{\pi}\int_{-\infty}^{\infty} |\hat{J}_2^{(l)}(z,\omega')|^2 h(\omega-\omega')d\omega' \Bigg] \end{aligned}$$

$$+\frac{\alpha}{s}\left[(Q_0 \log s - Q_1)|\hat{K}_1^{(1)}(z,\omega)|^2 - \frac{Q_0}{\pi}\int_{-\infty}^{\infty}|\hat{K}_2^{(1)}(z,\omega')|^2 h(\omega-\omega')d\omega'\right] \quad (8.152)$$

8.4.2.4 Case-IV: $\zeta_a = 1/4$

Here, the amplifier is placed at the boundary between the normal and anomalous GVD segments. The kernels are Eqs.(8.66) and (8.67). The only difference in this case from that of the previous one is that here the imaginary part of the kernels with opposite signs. But, again, it was shown in the previous subsection that the imaginary part does not make any contribution to the dynamics of quasi-linear pulses, and the sum of the spectral intensities is again conserved in this case during the pulse propagation.

References

1. F. Abdullaev, S. Darmanyan & P. Khabibullaev. *Optical Solitons.* Springer-Verlag, New York, NY. USA. (1993).
2. M. J. Ablowitz & H. Segur. *Solitons and the Inverse Scattering Transform.* SIAM. Philadelphia, PA. USA. (1981).
3. M. J. Ablowitz & G. Biondini. "Multiscale pulse dynamics in communication systems with strong dispersion management". *Optics Letters.* Vol 23, No 21, 1668-1670. (1998).
4. M. J. Ablowitz & T. Hirooka. "Resonant nonlinear interactions in strongly dispersion-managed transmission systems". *Optics Letters.* Vol 25, No 24, 1750-1752. (2000).
5. M. J. Ablowitz & T. Hirooka. "Nonlinear effects in quasi-linear dispersion-managed systems". *IEEE Photonics Technology Letters.* Vol 13, No 10, 1082-1084. (2001).
6. M. J. Ablowitz, T. Hirooka & G. Biondini. "Quasi-linear optical pulses in strongly dispersion-managed transmission system". *Optics Letters.* Vol 26, No 7, 459-461. (2001).
7. M. J. Ablowitz, G. Biondini & E. S. Olson. "Incomplete collisions of wavelength-division multiplexed dispersion-managed solitons". *Journal of Optical Society of America B.* Vol 18, No 3, 577-583. (2001).
8. M. J. Ablowitz & T. Hirooka. "Nonlinear effects in quasilinear dispersion-managed pulse transmission". *IEEE Journal of Photonics Technology Letters.* Vol 26, 1846-1848. (2001).
9. M. J. Ablowitz & T. Hirooka. "Intrachannel pulse interactions and timing shifts in strongly dispersion-managed transmission systems". *Optics Letters.* Vol 26, No 23, 1846-1848. (2001).
10. M. J. Ablowitz & T. Hirooka. "Intrachannel pulse interactions in dispersion-managed transmission systems: energy transfer". *Optics Letters.* Vol 27, No 3, 203-205. (2002).
11. M. J. Ablowitz & T. Hirooka. "Managing nonlinearity in strongly dispersion-managed optical pulse transmission". *Journal of Optical Society of America B.* Vol 19, No 3, 425-439. (2002).
12. M. J. Ablowitz, G. Biondini, A. Biswas, A. Docherty, T. Hirooka & S. Chakravarty. "Collision-induced timing shifts in dispersion-managed soliton systems". *Optics Letters.* Vol 27, 318-320. (2002).

13. C. D. Ahrens, M. J. Ablowitz, A. Docherty, V. Oleg, V. Sinkin, V. Gregorian & C. R. Menyuk. "Asymptotic analysis of collision-induced timing shifts in return-to-zero quasi-linear systems with pre- and post-dispersion compensation". *Optics Letters.* Vol 30, 2056-2058. (2005).
14. N. N. Akhmediev & A. Ankiewicz. *Solitons, Nonlinear Pulses and Beams*, Chapman and Hall, London. UK. (1997).
15. A. Biswas. "Dispersion-managed solitons in optical couplers". *Journal of Nonlinear Optical Physics and Materials.* Vol 12, No 1, 45-74. (2003).
16. A. Biswas. "Gabitov-Turitsyn equation for solitons in multiple channels". *Journal of Electromagnetic Waves and Applications.* Vol 17, No 11, 1539-1560. (2003).
17. A. Biswas. "Gabitov-Turitsyn equation for solitons in optical fibers". *Journal of Nonlinear Optical Physics and Materials.* Vol 12, No 1, 17-37. (2003).
18. A. Biswas. "Dispersion-Managed solitons in multiple channels". *Journal of Nonlinear Optical Physics and Materials.* Vol 13, No 1, 81-102. (2004).
19. A. Biswas. "Theory of quasi-linear pulses in optical fibers". *Optical Fiber Technology.* Vol 10, No 3, 232-259. (2004).
20. I. R. Gabitov & S. K. Turitsyn. "Average pulse dynamics in a cascaded transmission system with passive dispersion compensation". *Optics Letters.* Vol 21, No 5, 327-329. (1996).
21. I. R. Gabitov & S. K. Turitsyn. "Breathing solitons in optical fiber links". *JETP Letters.* Vol 63, No 10, 861-866. (1996).
22. I. R. Gabitov, E. G. Shapiro & S. K. Turitsyn. "Asymptotic breathing pulse in optical transmission systems with dispersion compensation". *Physical Review E.* Vol 55, No 3, 3624-3633. (1997).
23. I. R. Gabitov & P. M. Lushnikov. "Nonlinearity management in a dispersion compensation system". *Optics Letters.* Vol 27, No 2, 113-115. (2002).
24. I. R. Gabitov, R. Indik, L. Mollenauer, M. Shkarayev, M. Stepanov & P. M. Lushnikov. "Twin families of bisolitons in dispersion-managed systems". *Optics Letters.* Vol 32, No 6, 605-607. (2007.
25. A-H. Guan & Y-H. Wang. "Experimental study of interband and intraband crosstalk in WDM networks". *Optoelectronics Letters.* Vol 4, No 1, 42-44. (2008).
26. T. Hirooka & A. Hasegawa. "Chirped soliton interaction in strongly dispersion-managed wavelength-division-multiplexing systems". *Optics Letters.* Vol 23, No 10, 768-770. (1998).
27. T. Hirooka & S. Wabnitz. "Stabilization of dispersion-managed soliton transmission by nonlinear gain". *Electronics Letters.* Vol 35, No 8, 655-657. (1999).
28. T. Hirooka & S. Wabnitz. "Nonlinear gain control of dispersion-managed soliton amplitude and collisions". *Optical Fiber Technology.* Vol 6, No 2, 109-121. (2000).
29. V. E. Zakharov & S. Wabnitz. *Optical Solitons: Theoretical Challenges and Industrial Perspectives.* Springer, Heidelberg. DE. (1999).

Chapter 9
Higher Order Gabitov-Turitsyn Equations

9.1 Introduction

This chapter deals with the Gabitov-Turitsyn at the next higher order. This means that the multiple-scale expansion that was seen in Chapter 7, will now be carried out to $O(z_a^2)$. This gives a better accuracy and to the approximation that is described by the GTE seen in Chapter 7. In this technique, the pulse in the Fourier domain will be decomposed into a slowly evolving amplitude and a rapid phase that describe the chirp of the pulse. The fast phase is calculated explicitly that is driven by the large variations of the dispersion about the average. The amplitude evolution will be described by the nonlocal integro-differential evolution equations that is known as the higher order Gabitov-Turitsyn equation (HO-GTE).

The HO-GTE was first published by the remarkable work of Ablowitz et al. in 2002 [9]. This work was for the case of polarization preserving fibers. Subsequently, in 2007, Biswas extended this work to the cases of birefringent fibers and DWDM systems [14-16]. In this chapter, the section on polarization-preserving fibers is taken from Albowitz et al.'s work while the sections on birefringent fibers and DWDM systems are taken from Biswas's works. It needs to be noted that although the HO-GTE is useful for studying the structure and properties, it is, however, inconvenient for numerical computations because of the presence of the four-fold integrals that are given in the $O(z_a)$ terms of the HO-GTE, as will be seen.

9.2 Polarization preserving fibers

To obtain the asymptotic behavior for a polarization preserving fiber, the fast and slow z scales are introduced as

$$\zeta = \frac{z}{z_a} \qquad \text{and} \qquad Z = z \tag{9.1}$$

The field q is expanded in powers of z_a as

$$q(\zeta, Z, t) = q^{(0)}(\zeta, Z, t) + z_a q^{(1)}(\zeta, Z, t) + z_a^2 q^{(2)}(\zeta, Z, t) + \cdots \tag{9.2}$$

Equating coefficients of like powers of z_a gives

$$O(1/z_a): \quad i\frac{\partial q^{(0)}}{\partial \zeta} + \frac{\Delta(\zeta)}{2}\frac{\partial^2 q^{(0)}}{\partial t^2} = 0 \tag{9.3}$$

$$\begin{aligned} O(1): \quad & i\frac{\partial q^{(1)}}{\partial \zeta} + \frac{\Delta(\zeta)}{2}\frac{\partial^2 q^{(1)}}{\partial t^2} \\ & + \left\{ i\frac{\partial q^{(0)}}{\partial Z} + \frac{\delta_a}{2}\frac{\partial^2 q^{(0)}}{\partial t^2} + g(z)\left|q^{(0)}\right|^2 q^{(0)} \right\} = 0 \end{aligned} \tag{9.4}$$

$$\begin{aligned} O(z_a): \quad & i\frac{\partial q^{(2)}}{\partial \zeta} + \frac{\Delta(\zeta)}{2}\frac{\partial^2 q^{(2)}}{\partial t^2} + \left\{ i\frac{\partial q^{(1)}}{\partial Z} + \frac{\delta_a}{2}\frac{\partial^2 q^{(1)}}{\partial t^2} \right. \\ & \left. + \, g(z)\left[2|q^{(0)}|^2 q^{(1)} + (q^{(0)})^2 q^{(1)*}\right] \right\} = 0 \end{aligned} \tag{9.5}$$

Now, the Fourier transform and its inverse are respectively defined as

$$\hat{f}(\omega) = \int_{-\infty}^{\infty} f(t) e^{i\omega t} dt \tag{9.6}$$

$$f(t) = \frac{1}{2\pi}\int_{-\infty}^{\infty} \hat{f}(\omega) e^{-i\omega t} d\omega \tag{9.7}$$

At $O(1/z_a)$, Eq. (9.3), in the Fourier domain, is given by

$$i\frac{\partial \hat{q}^{(0)}}{\partial \zeta} - \frac{\omega^2}{2}\Delta(\zeta)\hat{q}^{(0)} = 0 \tag{9.8}$$

whose solution is

$$\hat{q}^{(0)}(\zeta, Z, \omega) = \hat{Q}^{(0)}(Z, \omega)\, e^{-\frac{i\omega^2}{2}C(\zeta)} \tag{9.9}$$

where

$$\hat{Q}^{(0)}(Z, \omega) = \hat{q}^{(0)}(0, Z, \omega) \tag{9.10}$$

and

$$C(\zeta) = \int_0^{\zeta} \Delta(\zeta')\, d\zeta' \tag{9.11}$$

At $O(1)$, Eq. (9.4) is solved in the Fourier domain by substituting the solution given by Eq.(9.9) into Eq.(9.4). This gives

$$i\frac{\partial \hat{q}^{(1)}}{\partial \zeta} - \frac{\omega^2}{2}\Delta(\zeta)\hat{q}^{(1)} = -e^{-\frac{i\omega^2}{2}C(\zeta)}\left(\frac{\partial \hat{Q}^{(0)}}{\partial Z} - \frac{\omega^2}{2}\delta_a \hat{Q}^{(0)}\right)$$

$$- g(\zeta)\int_{-\infty}^{\infty} |q^{(0)}|^2 q^{(0)} e^{i\omega t} dt \qquad (9.12)$$

Equation (9.12) is an inhomogeneous equations for $\hat{q}^{(1)}$ and with the homogeneous parts having the same structure as in Eq. (9.3). For the non-secularity condition of $\hat{q}^{(1)}$, it is necessary that the forcing terms are orthogonal to the adjoint solution of Eq.(9.3), a condition that is commonly known as Fredholm's Alternative (FA). This gives the condition for $\hat{Q}^{(0)}(Z,\omega)$ as

$$\frac{\partial \hat{Q}^{(0)}}{\partial Z} - \frac{\omega^2}{2}\delta_a \hat{Q}^{(0)} + \int_0^1 \int_{-\infty}^{\infty} e^{\frac{i\omega^2}{2}C(\zeta)} g(\zeta) |q^{(0)}|^2 q^{(0)} e^{i\omega t} dt d\zeta = 0 \qquad (9.13)$$

which can be simplified to

$$\frac{\partial \hat{Q}^{(0)}}{\partial Z} - \frac{\omega^2}{2}\delta_a \hat{Q}^{(0)} + \int_{-\infty}^{\infty}\int_{-\infty}^{\infty} r_0(\omega_1\omega_2)\, \hat{Q}^{(0)}(Z,\omega_1+\omega_2)$$

$$\cdot \hat{Q}^{(0)}(Z,\omega+\omega_1)\, \hat{Q}^{(0)}(Z,\omega+\omega_1+\omega_2)\, d\omega_1 d\omega_2 = 0 \qquad (9.14)$$

where, the kernel $r_0(x)$ is given by

$$r_0(x) = \frac{1}{(2\pi)^2}\int_0^1 g(\zeta) e^{ixC(\zeta)} d\zeta \qquad (9.15)$$

Equation (9.14) is the GTE for the propagation of solitons through polarization preserving fibers as seen in Chapter 7.

Equation (9.4) will now be solved to obtain $q^{(1)}(\zeta, Z, t)$. Substituting $\hat{Q}^{(0)}$ into the right-hand side of equation (9.12) and using Eqs.(9.9) and (9.13) give

$$\frac{\partial}{\partial \zeta}\left[i\hat{q}^{(1)} e^{\frac{i\omega^2}{2}C(\zeta)}\right] = \int_0^1 \int_{-\infty}^{\infty} g(\zeta) e^{\frac{i\omega^2}{2}C(\zeta)} |q^{(0)}|^2 q^{(0)} e^{i\omega t} dt d\zeta$$

$$- g(\zeta) e^{\frac{i\omega^2}{2}C(\zeta)} \int_{-\infty}^{\infty} |q^{(0)}|^2 q^{(0)} e^{i\omega t} dt \qquad (9.16)$$

which integrates to

$$i\hat{q}^{(1)} e^{\frac{i\omega^2}{2}C(\zeta)} = \hat{Q}^{(1)}(Z,\omega) + \zeta \int_0^1 \int_{-\infty}^{\infty} g(\zeta) e^{\frac{i\omega^2}{2}C(\zeta)} |q^{(0)}|^2 q^{(0)} e^{i\omega t} dt d\zeta$$

$$- \int_0^{\zeta} \int_{-\infty}^{\infty} g(\zeta') e^{\frac{i\omega^2}{2}C(\zeta')} |q^{(0)}|^2 q^{(0)} e^{i\omega t} dt d\zeta' \qquad (9.17)$$

where

$$\hat{Q}^{(1)}(Z,\omega) = i\hat{q}^{(1)}(0,Z,\omega)e^{\frac{i\omega^2}{2}C(0)} \tag{9.18}$$

Also, $\hat{Q}^{(1)}(Z,\omega)$ is so chosen that

$$\int_0^1 i\hat{q}^{(1)}e^{\frac{i\omega^2}{2}C(\zeta)}d\zeta = 0 \tag{9.19}$$

which is going to be an useful relation for subsequent orders. Applying Eq.(9.19) to Eq.(9.17) gives

$$\hat{Q}^{(1)}(Z,\omega) = \int_0^1\int_0^\zeta\int_{-\infty}^\infty g(\zeta')e^{\frac{i\omega^2}{2}C(\zeta')}|q^{(0)}|^2q^{(0)}e^{i\omega t}dtd\zeta'd\zeta$$
$$-\frac{1}{2}\int_0^1\int_{-\infty}^\infty g(\zeta)e^{\frac{i\omega^2}{2}C(\zeta)}|q^{(0)}|^2q^{(0)}e^{i\omega t}dtd\zeta \tag{9.20}$$

Now, Eq.(9.17) by virtue of Eq.(9.20) can be written as

$$\hat{q}^{(1)}(\zeta,Z,\omega) = ie^{\frac{i\omega^2}{2}C(\zeta)}\left[\int_0^\zeta\int_{-\infty}^\infty g(\zeta')e^{\frac{i\omega^2}{2}C(\zeta')}|q^{(0)}|^2q^{(0)}e^{i\omega t}dtd\zeta'\right.$$
$$-\int_0^1\int_0^\zeta\int_{-\infty}^\infty g(\zeta')e^{\frac{i\omega^2}{2}C(\zeta')}|q^{(0)}|^2q^{(0)}e^{i\omega t}dtd\zeta'd\zeta$$
$$\left.-\left(\zeta-\frac{1}{2}\right)\int_0^1\int_{-\infty}^\infty g(\zeta)e^{\frac{i\omega^2}{2}C(\zeta)}|q^{(0)}|^2q^{(0)}e^{i\omega t}dtd\zeta\right] \tag{9.21}$$

which can also be rewritten as

$$\hat{q}^{(1)}(\zeta,Z,\omega) = ie^{-\frac{i\omega^2}{2}C(\zeta)}\left[\int_{-\infty}^\infty\int_{-\infty}^\infty \hat{Q}^{(0)*}(\omega+\Omega_1+\Omega_2)\right.$$
$$\cdot\hat{Q}^{(0)}(\omega+\Omega_1)\hat{Q}^{(0)}(\omega+\Omega_2)$$
$$\cdot\left\{\int_0^\zeta g(\zeta')e^{i\Omega_1\Omega_2C(\zeta')}d\zeta' - \int_0^1\int_0^\zeta g(\zeta')e^{i\Omega_1\Omega_2C(\zeta')}d\zeta'd\zeta\right.$$
$$\left.\left.-\left(\zeta-\frac{1}{2}\right)\int_0^1 g(\zeta)e^{i\Omega_1\Omega_2C(\zeta)}d\zeta\right\}d\Omega_1\Omega_2\right] \tag{9.22}$$

Thus, at $O(z_a)$,

$$\hat{q}(\zeta,Z,\omega) = \hat{q}^{(0)}(\zeta,Z,\omega) + z_a\hat{q}^{(1)}(\zeta,Z,\omega) \tag{9.23}$$

Moving on to the next order at $O(z_a^2)$, one can note that the GTE given by Eq. (9.14) is allowed to have an additional term of $O(z_a)$ as

$$\frac{\partial \hat{Q}^{(0)}}{\partial Z} - \frac{\omega^2}{2}\delta_a \hat{Q}^{(0)} + \int_{-\infty}^{\infty}\int_{-\infty}^{\infty} r_0\left(\omega_1\omega_2\right)\hat{Q}^{(0)}\left(Z,\omega_1+\omega_2\right)$$
$$\cdot\hat{Q}^{(0)}\left(Z,\omega+\omega_1\right)\hat{Q}^{(0)}\left(Z,\omega+\omega_1+\omega_2\right)d\omega_1 d\omega_2$$
$$= z_a\hat{n}(Z,\omega) + O(z_a^2) \tag{9.24}$$

The higher order correction $\hat{n}$ can be obtained from the suitable non-secular conditions at $O(z_a^2)$ in Eq.(9.14). Now, Eq. (9.5), in the Fourier domain, is

$$\frac{\partial}{\partial \zeta}\left[i\hat{q}^{(2)}e^{\frac{i\omega^2}{2}C(\zeta)}\right] + \hat{n} + e^{\frac{i\omega^2}{2}C(\zeta)}\left(i\frac{\partial \hat{q}^{(1)}}{\partial Z} - \frac{\omega^2}{2}\delta_a\hat{q}^{(1)}\right)$$
$$+ e^{\frac{i\omega^2}{2}C(\zeta)}g(\zeta)\int_{-\infty}^{\infty}\left[2|q^{(0)}|^2 q^{(1)} + (q^{(0)})^2 q^{(1)*}\right]e^{i\omega t}dt = 0 \tag{9.25}$$

But, again, Eq.(9.19) gives

$$\int_0^1 \hat{q}^{(1)} e^{\frac{i\omega^2}{2}C(\zeta)}d\zeta = 0 \tag{9.26}$$

Applying the non-secularity condition (9.26) to Eq.(9.25) gives

$$\hat{n} = -\int_0^1\int_{-\infty}^{\infty} e^{\frac{i\omega^2}{2}C(\zeta)}g(\zeta)\left[2|q^{(0)}|^2 q^{(1)} + (q^{(0)})^2 q^{(1)*}\right]e^{i\omega t}dt d\zeta \tag{9.27}$$

Using Eqs.(9.9) and (9.21),Eq.(9.27) can be rewritten as

$$\hat{n} = \int_{-\infty}^{\infty}\int_{-\infty}^{\infty}\int_{-\infty}^{\infty}\int_{-\infty}^{\infty} r_1\left(\omega_1\omega_2, \Omega_1\Omega_2\right)$$
$$\cdot\Big[2\hat{Q}^{(0)}\left(\omega+\omega_1\right)\hat{Q}^{(0)*}\left(\omega+\omega_1+\omega_2\right)\hat{Q}^{(0)}\left(\omega+\omega_2+\Omega_1\right)$$
$$\cdot\hat{Q}^{0)}\left(\omega+\omega_2+\Omega_2\right)\hat{Q}^{(0)*}\left(\omega+\omega_2+\Omega_1+\Omega_2\right)$$
$$-\hat{Q}^{(0)}\left(\omega+\omega_1\right)\hat{Q}^{(0)}\left(\omega+\omega_2\right)\hat{Q}^{(0)*}\left(\omega+\omega_1+\omega_2+\Omega_1\right)$$
$$\cdot\hat{Q}^{(0)*}\left(\omega+\omega_1+\omega_2-\Omega_2\right)$$
$$\cdot\hat{Q}^{(0)*}\left(\omega+\omega_1+\omega_2+\Omega_1-\Omega_2\right)\Big]d\omega_1 d\omega_2 d\Omega_1 d\Omega_2 \tag{9.28}$$

where, the kernel $r_1(x,y)$ is given by

$$r_1(x,y) = \frac{1}{(2\pi)^4}\int_0^1\int_0^\zeta g(\zeta)g(\zeta')e^{i(xC(\zeta)+yC(\zeta))}d\zeta d\zeta'$$
$$-\left[\int_0^1 g(\zeta)e^{ixC(\zeta)}d\zeta\right]\left[\int_0^\zeta g(\zeta')e^{ixC(\zeta')}d\zeta'\right]$$
$$-\left[\int_0^1\left(\zeta-\frac{1}{2}\right)g(\zeta)e^{ixC(\zeta)}d\zeta\right]\left[\int_0^1 g(\zeta)e^{iyC(\zeta)}d\zeta\right] \tag{9.29}$$

Equation (9.24) represents the HO-GTE for the propagation of solitons through polarization preserving optical fibers.

9.3 Birefringent fibers

The equations that describe the pulse propagation in birefringent fibers are of the following dimensionless form:

$$iu_z + \frac{D(z)}{2}u_{tt} + g(z)(|u|^2 + \alpha|v|^2)u = 0 \tag{9.30}$$

$$iv_z + \frac{D(z)}{2}v_{tt} + g(z)(|v|^2 + \alpha|u|^2)v = 0 \tag{9.31}$$

where α represents the XPM coefficients. The fields u and v are expanded in powers of z_a as

$$u(\zeta, Z, t) = u^{(0)}(\zeta, Z, t) + z_a u^{(1)}(\zeta, Z, t) + z_a^2 u^{(2)}(\zeta, Z, t) + \cdots \tag{9.32}$$

and

$$v(\zeta, Z, t) = v^{(0)}(\zeta, Z, t) + z_a v^{(1)}(\zeta, Z, t) + z_a^2 v^{(2)}(\zeta, Z, t) + \cdots \tag{9.33}$$

Equating coefficients of like powers of z_a gives

$$O\left(\frac{1}{z_a}\right): \quad i\frac{\partial u^{(0)}}{\partial \zeta} + \frac{\Delta(\zeta)}{2}\frac{\partial^2 u^{(0)}}{\partial t^2} = 0 \tag{9.34}$$

$$O\left(\frac{1}{z_a}\right): \quad i\frac{\partial v^{(0)}}{\partial \zeta} + \frac{\Delta(\zeta)}{2}\frac{\partial^2 v^{(0)}}{\partial t^2} = 0 \tag{9.35}$$

$$O(1): \quad i\frac{\partial u^{(1)}}{\partial \zeta} + \frac{\Delta(\zeta)}{2}\frac{\partial^2 u^{(1)}}{\partial t^2} + \left\{ i\frac{\partial u^{(0)}}{\partial Z} + \frac{\delta_a}{2}\frac{\partial^2 u^{(0)}}{\partial t^2} \right.$$
$$\left. + \, g(z)\left(|u^{(0)}|^2 + \alpha|v^{(0)}|^2\right)u^{(0)} \right\} = 0 \tag{9.36}$$

$$O(1): \quad i\frac{\partial v^{(1)}}{\partial \zeta} + \frac{\Delta(\zeta)}{2}\frac{\partial^2 v^{(1)}}{\partial t^2} + \left\{ i\frac{\partial v^{(0)}}{\partial Z} + \frac{\delta_a}{2}\frac{\partial^2 v^{(0)}}{\partial t^2} \right.$$
$$\left. + \, g(z)\left(|v^{(0)}|^2 + \alpha|u^{(0)}|^2\right)v^{(0)} \right\} = 0 \tag{9.37}$$

$$O(z_a): \quad i\frac{\partial u^{(2)}}{\partial \zeta} + \frac{\Delta(\zeta)}{2}\frac{\partial^2 u^{(2)}}{\partial t^2} + \left\{ i\frac{\partial u^{(1)}}{\partial Z} + \frac{\delta_a}{2}\frac{\partial^2 u^{(1)}}{\partial t^2} + g(z) \right.$$

$$\cdot\left[2|u^{(0)}|^2u^{(1)}+(u^{(0)})^2u^{(1)*}+\alpha\{2|v^{(0)}|^2v^{(1)}+(v^{(0)})^2v^{(1)*}\}\right]\Bigg\} \tag{9.38}$$

$$O(z_a):\quad i\frac{\partial v^{(2)}}{\partial\zeta}+\frac{\Delta(\zeta)}{2}\frac{\partial^2v^{(2)}}{\partial t^2}+\Bigg\{i\frac{\partial v^{(1)}}{\partial Z}+\frac{\delta_a}{2}\frac{\partial^2v^{(1)}}{\partial t^2}+g(z)$$

$$\cdot\left[2|v^{(0)}|^2v^{(1)}+(v^{(0)})^2v^{(1)*}+\alpha\{2|u^{(0)}|^2u^{(1)}+(u^{(0)})^2u^{(1)*}\}\right]\Bigg\} \tag{9.39}$$

At $O(1/z_a)$, Eqs.(9.34) and (9.35), respectively, in the Fourier domain are given by

$$i\frac{\partial\hat{u}^{(0)}}{\partial\zeta}-\frac{\omega^2}{2}\Delta(\zeta)\hat{u}^{(0)}=0 \tag{9.40}$$

and

$$i\frac{\partial\hat{v}^{(0)}}{\partial\zeta}-\frac{\omega^2}{2}\Delta(\zeta)\hat{v}^{(0)}=0 \tag{9.41}$$

whose respective solutions are

$$\hat{u}^{(0)}\left(\zeta,Z,\omega\right)=\hat{U}_0\left(Z,\omega\right)e^{-\frac{i\omega^2}{2}C(\zeta)} \tag{9.42}$$

and

$$\hat{v}^{(0)}\left(\zeta,Z,\omega\right)=\hat{V}_0\left(Z,\omega\right)e^{-\frac{i\omega^2}{2}C(\zeta)} \tag{9.43}$$

where

$$\hat{U}_0\left(Z,\omega\right)=\hat{u}^{(0)}\left(0,Z,\omega\right) \tag{9.44}$$

and

$$\hat{V}_0\left(Z,\omega\right)=\hat{v}^{(0)}\left(0,Z,\omega\right) \tag{9.45}$$

At $O(1)$, Eqs. (9.36) and (9.37) are solved in the Fourier domain by substituting the respective solutions given by Eqs.(9.42) and (9.43) into Eqs.(9.36) and (9.37). This gives

$$i\frac{\partial\hat{u}^{(1)}}{\partial\zeta}-\frac{\omega^2}{2}\Delta(\zeta)\hat{u}^{(1)}=-e^{-\frac{i\omega^2}{2}C(\zeta)}\left(\frac{\partial\hat{U}_0}{\partial Z}-\frac{\omega^2}{2}\delta_a\hat{U}_0\right)-g(\zeta)$$

$$\cdot\int_{-\infty}^{\infty}\left(|u^{(0)}|^2+\alpha|v^{(0)}|^2\right)u^{(0)}e^{i\omega t}dt \tag{9.46}$$

and

$$i\frac{\partial\hat{v}^{(1)}}{\partial\zeta}-\frac{\omega^2}{2}\Delta(\zeta)\hat{v}^{(1)}=-e^{-\frac{i\omega^2}{2}C(\zeta)}\left(\frac{\partial\hat{V}_0}{\partial Z}-\frac{\omega^2}{2}\delta_a\hat{V}_0\right)-g(\zeta)$$

$$\cdot\int_{-\infty}^{\infty}\left(|v^{(0)}|^2+\alpha|u^{(0)}|^2\right)v^{(0)}e^{i\omega t}dt \tag{9.47}$$

Equations (9.46) and (9.47) are inhomogeneous equations for $\hat{u}^{(1)}$ and $\hat{v}^{(1)}$, respectively, with the homogeneous parts having the same structures as in Eqs.(9.34) and (9.35), respectively. For the non-secularity conditions of $\hat{u}^{(1)}$ and $\hat{v}^{(1)}$, FA gives the conditions for $\hat{U}_0(Z,\omega)$ and $\hat{V}_0(Z,\omega)$ respectively as

$$\frac{\partial \hat{U}_0}{\partial Z} - \frac{\omega^2}{2}\delta_a \hat{U}_0 + \int_0^1 \int_{-\infty}^{\infty} e^{\frac{i\omega^2}{2}C(\zeta)} g(\zeta) \cdot \left(|u^{(0)}|^2 + \alpha |v^{(0)}|^2\right) u^{(0)} e^{i\omega t} dt d\zeta = 0 \tag{9.48}$$

and

$$\frac{\partial \hat{V}_0}{\partial Z} - \frac{\omega^2}{2}\delta_a \hat{V}_0 + \int_0^1 \int_{-\infty}^{\infty} e^{\frac{i\omega^2}{2}C(\zeta)} g(\zeta) \cdot \left(|v^{(0)}|^2 + \alpha |u^{(0)}|^2\right) v^{(0)} e^{i\omega t} dt d\zeta = 0 \tag{9.49}$$

Equations (9.48) and (9.49) can be respectively simplified to

$$\begin{aligned}\frac{\partial \hat{U}_0}{\partial Z} - \frac{\omega^2}{2}\delta_a \hat{U}_0 + \int_{-\infty}^{\infty}\int_{-\infty}^{\infty} r_0(\omega_1\omega_2)\, \hat{U}_0(Z,\omega_1+\omega_2) \\ \cdot \Big[\hat{U}_0(Z,\omega+\omega_1)\, \hat{U}_0(Z,\omega+\omega_1+\omega_2) \\ + \alpha \hat{V}_0(Z,\omega+\omega_1)\, \hat{V}_0(Z,\omega+\omega_1+\omega_2)\Big] d\omega_1 d\omega_2 = 0\end{aligned} \tag{9.50}$$

and

$$\begin{aligned}\frac{\partial \hat{V}_0}{\partial Z} - \frac{\omega^2}{2}\delta_a \hat{V}_0 + \int_{-\infty}^{\infty}\int_{-\infty}^{\infty} r_0(\omega_1\omega_2)\, \hat{V}_0(Z,\omega_1+\omega_2) \\ \cdot \Big[\hat{V}_0(Z,\omega+\omega_1)\, \hat{V}_0(Z,\omega+\omega_1+\omega_2) \\ + \alpha \hat{U}_0(Z,\omega+\omega_1)\, \hat{U}_0(Z,\omega+\omega_1+\omega_2)\Big] d\omega_1 d\omega_2 = 0\end{aligned} \tag{9.51}$$

Equations (9.50) and (9.51) are the GTE for the propagation of solitons through a birefringent fiber as seen in Chapter 7. Equations (9.46) and (9.47) will now be solved to obtain $u^{(1)}(\zeta,Z,t)$ and $v^{(1)}(\zeta,Z,t)$. Substituting $\hat{U}_0$ and $\hat{V}_0$ into the right-hand side of Eqs. (9.46) and (9.47), respectively, and using pairs (9.42)–(9.43) and (9.48)–(9.49) give

$$\begin{aligned}&\frac{\partial}{\partial \zeta}\left[i\hat{u}^{(1)} e^{\frac{i\omega^2}{2}C(\zeta)}\right] \\ &= \int_0^1\int_{-\infty}^{\infty} g(\zeta) e^{\frac{i\omega^2}{2}C(\zeta)} \left(|u^{(0)}|^2 + \alpha |v^{(0)}|^2\right) u^{(0)} e^{i\omega t} dt d\zeta \\ &\quad - g(\zeta) e^{\frac{i\omega^2}{2}C(\zeta)} \int_{-\infty}^{\infty} \left(|u^{(0)}|^2 + \alpha |v^{(0)}|^2\right) u^{(0)} e^{i\omega t} dt\end{aligned} \tag{9.52}$$

and

$$\frac{\partial}{\partial\zeta}\left[i\hat{v}^{(1)}e^{\frac{i\omega^2}{2}C(\zeta)}\right]$$

$$= \int_0^1\int_{-\infty}^{\infty} g(\zeta)e^{\frac{i\omega^2}{2}C(\zeta)}\left(|v^{(0)}|^2 + \alpha|u^{(0)}|^2\right)v^{(0)}e^{i\omega t}dtd\zeta$$

$$- g(\zeta)e^{\frac{i\omega^2}{2}C(\zeta)}\int_{-\infty}^{\infty}\left(|v^{(0)}|^2 + \alpha|u^{(0)}|^2\right)v^{(0)}e^{i\omega t}dt \tag{9.53}$$

Integration of Eqs. (9.52) and (9.53) yields

$$i\hat{u}^{(1)}e^{\frac{i\omega^2}{2}C(\zeta)}$$

$$= \hat{U}_1(Z,\omega) + \zeta\int_0^1\int_{-\infty}^{\infty} g(\zeta)e^{\frac{i\omega^2}{2}C(\zeta)}\left(|u^{(0)}|^2 + \alpha|v^{(0)}|^2\right)u^{(0)}e^{i\omega t}dtd\zeta$$

$$- \int_0^{\zeta}\int_{-\infty}^{\infty} g(\zeta')e^{\frac{i\omega^2}{2}C(\zeta')}\left(|u^{(0)}|^2 + \alpha|v^{(0)}|^2\right)u^{(0)}e^{i\omega t}dtd\zeta' \tag{9.54}$$

and

$$i\hat{v}^{(1)}e^{\frac{i\omega^2}{2}C(\zeta)}$$

$$= \hat{V}_1(Z,\omega) + \zeta\int_0^1\int_{-\infty}^{\infty} g(\zeta)e^{\frac{i\omega^2}{2}C(\zeta)}\left(|v^{(0)}|^2 + \alpha|u^{(0)}|^2\right)v^{(0)}e^{i\omega t}dtd\zeta$$

$$- \int_0^{\zeta}\int_{-\infty}^{\infty} g(\zeta')e^{\frac{i\omega^2}{2}C(\zeta')}\left(|v^{(0)}|^2 + \alpha|u^{(0)}|^2\right)v^{(0)}e^{i\omega t}dtd\zeta' \tag{9.55}$$

where

$$\hat{U}_1(Z,\omega) = i\hat{u}^{(1)}(0,Z,\omega)e^{\frac{i\omega^2}{2}C(0)} \tag{9.56}$$

and

$$\hat{V}_1(Z,\omega) = i\hat{v}^{(1)}(0,Z,\omega)e^{\frac{i\omega^2}{2}C(0)} \tag{9.57}$$

Also, $\hat{U}_1(Z,\omega)$ and $\hat{V}_1(Z,\omega)$ are so chosen that

$$\int_0^1 i\hat{u}^{(1)}e^{\frac{i\omega^2}{2}C(\zeta)}d\zeta = 0 \tag{9.58}$$

and

$$\int_0^1 i\hat{v}^{(1)}e^{\frac{i\omega^2}{2}C(\zeta)}d\zeta = 0 \tag{9.59}$$

which are going to be useful relations at subsequent orders. Applying Eqs.(9.58) and (9.59) to Eqs.(9.54) and (9.55), respectively, yields

$$\hat{U}_1(Z,\omega) = \int_0^1 \int_0^\zeta \int_{-\infty}^\infty g(\zeta')e^{\frac{i\omega^2}{2}C(\zeta')} \left(|u^{(0)}|^2 + \alpha|v^{(0)}|^2\right) u^{(0)} e^{i\omega t} dt d\zeta' d\zeta$$

$$-\frac{1}{2}\int_0^1 \int_{-\infty}^\infty g(\zeta)e^{\frac{i\omega^2}{2}C(\zeta)} \left(|u^{(0)}|^2 + \alpha|v^{(0)}|^2\right) u^{(0)} e^{i\omega t} dt d\zeta \quad (9.60)$$

and

$$\hat{V}_1(Z,\omega) = \int_0^1 \int_0^\zeta \int_{-\infty}^\infty g(\zeta')e^{\frac{i\omega^2}{2}C(\zeta')} \left(|v^{(0)}|^2 + \alpha|u^{(0)}|^2\right) v^{(0)} e^{i\omega t} dt d\zeta' d\zeta$$

$$-\frac{1}{2}\int_0^1 \int_{-\infty}^\infty g(\zeta)e^{\frac{i\omega^2}{2}C(\zeta)} \left(|v^{(0)}|^2 + \alpha|u^{(0)}|^2\right) v^{(0)} e^{i\omega t} dt d\zeta \quad (9.61)$$

Now, Eqs. (9.54) and (9.55), by virtue of Eqs.(9.60) and (9.61), can be respectively written as

$$\begin{aligned}
&\hat{u}^{(1)}(\zeta,Z,\omega) \\
&= ie^{\frac{i\omega^2}{2}C(\zeta)}\left[\int_0^\zeta \int_{-\infty}^\infty g(\zeta')e^{\frac{i\omega^2}{2}C(\zeta')} \left(|u^{(0)}|^2 + \alpha|v^{(0)}|^2\right) u^{(0)} e^{i\omega t} dt d\zeta'\right. \\
&\quad - \int_0^1 \int_0^\zeta \int_{-\infty}^\infty g(\zeta')e^{\frac{i\omega^2}{2}C(\zeta')} \left(|u^{(0)}|^2 + \alpha|v^{(0)}|^2\right) u^{(0)} e^{i\omega t} dt d\zeta' d\zeta \\
&\quad \left. - \left(\zeta - \frac{1}{2}\right)\int_0^1 \int_{-\infty}^\infty g(\zeta)e^{\frac{i\omega^2}{2}C(\zeta)} \left(|u^{(0)}|^2 + \alpha|v^{(0)}|^2\right) u^{(0)} e^{i\omega t} dt d\zeta\right]
\end{aligned} \quad (9.62)$$

and

$$\begin{aligned}
&\hat{v}^{(1)}(\zeta,Z,\omega) \\
&= ie^{\frac{i\omega^2}{2}C(\zeta)}\left[\int_0^\zeta \int_{-\infty}^\infty g(\zeta')e^{\frac{i\omega^2}{2}C(\zeta')} \left(|v^{(0)}|^2 + \alpha|u^{(0)}|^2\right) v^{(0)} e^{i\omega t} dt d\zeta'\right. \\
&\quad - \int_0^1 \int_0^\zeta \int_{-\infty}^\infty g(\zeta')e^{\frac{i\omega^2}{2}C(\zeta')} \left(|v^{(0)}|^2 + \alpha|u^{(0)}|^2\right) v^{(0)} e^{i\omega t} dt d\zeta' d\zeta \\
&\quad \left. - \left(\zeta - \frac{1}{2}\right)\int_0^1 \int_{-\infty}^\infty g(\zeta)e^{\frac{i\omega^2}{2}C(\zeta)} \left(|v^{(0)}|^2 + \alpha|u^{(0)}|^2\right) v^{(0)} e^{i\omega t} dt d\zeta\right]
\end{aligned} \quad (9.63)$$

Equations (9.62) and (9.63) can now be respectively rewritten as

$$\hat{u}^{(1)}(\zeta,Z,\omega) = ie^{-\frac{i\omega^2}{2}C(\zeta)}\left[\int_{-\infty}^\infty \int_{-\infty}^\infty \hat{U}_0^*\left(\omega+\Omega_1+\Omega_2\right)\left\{\hat{U}_0\left(\omega+\Omega_1\right)\right.\right.$$
$$\left.\cdot\hat{U}_0\left(\omega+\Omega_2\right) + \alpha\hat{V}_0\left(\omega+\Omega_1\right)\hat{V}_0\left(\omega+\Omega_2\right)\right\}$$

$$\cdot\left\{\int_0^{\zeta} g(\zeta')e^{i\Omega_1\Omega_2 C(\zeta')}d\zeta' - \int_0^1\int_0^{\zeta} g(\zeta')e^{i\Omega_1\Omega_2 C(\zeta')}d\zeta' d\zeta\right.$$
$$\left. - \left(\zeta - \frac{1}{2}\right)\int_0^1 g(\zeta)e^{i\Omega_1\Omega_2 C(\zeta)}d\zeta\right\}d\Omega_1\Omega_2\Bigg] \tag{9.64}$$

and

$$\hat{v}^{(1)}(\zeta, Z, \omega) = ie^{-\frac{i\omega^2}{2}C(\zeta)}\left[\int_{-\infty}^{\infty}\int_{-\infty}^{\infty} \hat{V}_0^*\left(\omega + \Omega_1 + \Omega_2\right)\left\{\hat{V}_0\left(\omega + \Omega_1\right)\right.\right.$$
$$\left.\cdot\hat{V}_0\left(\omega + \Omega_2\right) + \alpha\hat{U}_0\left(\omega + \Omega_1\right)\hat{U}_0\left(\omega + \Omega_2\right)\right\}$$
$$\cdot\left\{\int_0^{\zeta} g(\zeta')e^{i\Omega_1\Omega_2 C(\zeta')}d\zeta' - \int_0^1\int_0^{\zeta} g(\zeta')e^{i\Omega_1\Omega_2 C(\zeta')}d\zeta' d\zeta\right.$$
$$\left. - \left(\zeta - \frac{1}{2}\right)\int_0^1 g(\zeta)e^{i\Omega_1\Omega_2 C(\zeta)}d\zeta\right\}d\Omega_1\Omega_2\Bigg] \tag{9.65}$$

Thus, at $O(z_a)$,

$$\hat{u}(\zeta, Z, \omega) = \hat{u}^{(0)}(\zeta, Z, \omega) + z_a\hat{u}^{(1)}(\zeta, Z, \omega) \tag{9.66}$$

and

$$\hat{v}(\zeta, Z, \omega) = \hat{v}^{(0)}(\zeta, Z, \omega) + z_a\hat{v}^{(1)}(\zeta, Z, \omega) \tag{9.67}$$

Moving on to the next order at $O(z_a^2)$, one can note that the GTE given by Eqs. (9.50) and (9.51) are allowed to have an additional term of $O(z_a)$ such as

$$\frac{\partial\hat{U}_0}{\partial Z} - \frac{\omega^2}{2}\delta_a\hat{U}_0 + \int_{-\infty}^{\infty}\int_{-\infty}^{\infty} r_0\left(\omega_1\omega_2\right)\hat{U}_0\left(Z, \omega_1 + \omega_2\right)$$
$$\cdot[\hat{U}_0\left(Z, \omega + \omega_1\right)\hat{U}_0\left(Z, \omega + \omega_1 + \omega_2\right) + \alpha\hat{V}_0\left(Z, \omega + \omega_1\right)$$
$$\cdot\hat{V}_0\left(Z, \omega + \omega_1 + \omega_2\right)]d\omega_1 d\omega_2 = z_a\hat{n}_1(Z, \omega) + O(z_a^2) \tag{9.68}$$

and

$$\frac{\partial\hat{V}_0}{\partial Z} - \frac{\omega^2}{2}\delta_a\hat{V}_0 + \int_{-\infty}^{\infty}\int_{-\infty}^{\infty} r_0\left(\omega_1\omega_2\right)\hat{V}_0\left(Z, \omega_1 + \omega_2\right)$$
$$\cdot[\hat{V}_0\left(Z, \omega + \omega_1\right)\hat{V}_0\left(Z, \omega + \omega_1 + \omega_2\right) + \alpha\hat{U}_0\left(Z, \omega + \omega_1\right)$$
$$\cdot\hat{U}_0\left(Z, \omega + \omega_1 + \omega_2\right)]d\omega_1 d\omega_2 = z_a\hat{n}_2(Z, \omega) + O(z_a^2) \tag{9.69}$$

The higher order corrections $\hat{n}_1$ and $\hat{n}_2$ can be obtained from suitable non-secular conditions at $O(z_a^2)$ in Eq. (9.38) and Eq.(9.39), respectively. Now, Eqs. (9.38) and (9.39), in the Fourier domain, respectively, are

$$\frac{\partial}{\partial\zeta}\left[i\hat{u}^{(2)}e^{\frac{i\omega^2}{2}C(\zeta)}\right] + \hat{n}_1 + e^{\frac{i\omega^2}{2}C(\zeta)}\left(i\frac{\partial\hat{u}^{(1)}}{\partial Z} - \frac{\omega^2}{2}\delta_a\hat{u}^{(1)}\right)$$

$$+ e^{\frac{i\omega^2}{2}C(\zeta)} g(\zeta) \int_{-\infty}^{\infty} \left[2|u^{(0)}|^2 u^{(1)} + (u^{(0)})^2 u^{(1)*} \right.$$
$$\left. + \alpha \left\{ 2|v^{(0)}|^2 v^{(1)} + (v^{(0)})^2 v^{(1)*} \right\} \right] e^{i\omega t} dt = 0 \tag{9.70}$$

and

$$\frac{\partial}{\partial \zeta} \left[i\hat{v}^{(2)} e^{\frac{i\omega^2}{2}C(\zeta)} \right] + \hat{n}_2 + e^{\frac{i\omega^2}{2}C(\zeta)} \left(i\frac{\partial \hat{v}^{(1)}}{\partial Z} - \frac{\omega^2}{2}\delta_a \hat{v}^{(1)} \right)$$
$$+ e^{\frac{i\omega^2}{2}C(\zeta)} g(\zeta) \int_{-\infty}^{\infty} \left[2|v^{(0)}|^2 v^{(1)} + (v^{(0)})^2 v^{(1)*} \right.$$
$$\left. + \alpha \left\{ 2|u^{(0)}|^2 u^{(1)} + (u^{(0)})^2 u^{(1)*} \right\} \right] e^{i\omega t} dt = 0 \tag{9.71}$$

But, again, Eqs.(9.58) and (9.59) give

$$\int_0^1 \hat{u}^{(1)} e^{\frac{i\omega^2}{2}C(\zeta)} d\zeta = 0 \tag{9.72}$$

and

$$\int_0^1 \hat{v}^{(1)} e^{\frac{i\omega^2}{2}C(\zeta)} d\zeta = 0 \tag{9.73}$$

Applying the non-secularity conditions (9.72) and (9.73) to Eq.(9.70) and Eq.(9.71), respectively, gives

$$\hat{n}_1 = -\int_0^1 \int_{-\infty}^{\infty} e^{\frac{i\omega^2}{2}C(\zeta)} g(\zeta) \left[2|u^{(0)}|^2 u^{(1)} + (u^{(0)})^2 u^{(1)*} \right.$$
$$\left. + \alpha \left\{ 2|v^{(0)}|^2 v^{(1)} + (v^{(0)})^2 v^{(1)*} \right\} \right] e^{i\omega t} dt d\zeta \tag{9.74}$$

and

$$\hat{n}_2 = -\int_0^1 \int_{-\infty}^{\infty} e^{\frac{i\omega^2}{2}C(\zeta)} g(\zeta) \left[2|v^{(0)}|^2 v^{(1)} + (v^{(0)})^2 v^{(1)*} \right.$$
$$\left. + \alpha \left\{ 2|u^{(0)}|^2 u^{(1)} + (u^{(0)})^2 u^{(1)*} \right\} \right] e^{i\omega t} dt d\zeta \tag{9.75}$$

Using pairs (9.42)–(9.43) and (9.62)–(9.63), Eqs.(9.74) and (9.75) can respectively be written as

$$\hat{n}_1 = \int_{-\infty}^{\infty} \int_{-\infty}^{\infty} \int_{-\infty}^{\infty} \int_{-\infty}^{\infty} r_1 \left(\omega_1 \omega_2, \Omega_1 \Omega_2 \right)$$
$$\cdot \left[\left\{ 2\hat{U}_0 \left(\omega + \omega_1 \right) \hat{U}_0^* \left(\omega + \omega_1 + \omega_2 \right) \hat{U}_0 \left(\omega + \omega_2 + \Omega_1 \right) \right. \right.$$
$$\cdot \hat{U}_0 \left(\omega + \omega_2 + \Omega_2 \right) \hat{U}_0^* \left(\omega + \omega_2 + \Omega_1 + \Omega_2 \right)$$
$$- \hat{U}_0 \left(\omega + \omega_1 \right) \hat{U}_0 \left(\omega + \omega_2 \right) \hat{U}_0^* \left(\omega + \omega_1 + \omega_2 + \Omega_1 \right)$$

$$
\begin{aligned}
&\cdot \hat{U}_0^*\left(\omega+\omega_1+\omega_2-\Omega_2\right)\hat{U}_0^*\left(\omega+\omega_1+\omega_2+\Omega_1-\Omega_2\right)\Big\} \\
&+\alpha\left\{2\hat{V}_0\left(\omega+\omega_1\right)\hat{V}_0^*\left(\omega+\omega_1+\omega_2\right)\hat{V}_0\left(\omega+\omega_2+\Omega_1\right)\right. \\
&\cdot \hat{V}_0\left(\omega+\omega_2+\Omega_2\right)\hat{V}_0^*\left(\omega+\omega_2+\Omega_1+\Omega_2\right) \\
&-\hat{V}_0\left(\omega+\omega_1\right)\hat{V}_0\left(\omega+\omega_2\right)\hat{V}_0^*\left(\omega+\omega_1+\omega_2+\Omega_1\right) \\
&\cdot\hat{V}_0^*\left(\omega+\omega_1+\omega_2-\Omega_2\right) \\
&\left.\left.\cdot \hat{V}_0^*\left(\omega+\omega_1+\omega_2+\Omega_1-\Omega_2\right)\right\}\right] d\omega_1 d\omega_2 d\Omega_1 d\Omega_2
\end{aligned}
\tag{9.76}
$$

and

$$
\begin{aligned}
\hat{n}_2 =& \int_{-\infty}^{\infty}\int_{-\infty}^{\infty}\int_{-\infty}^{\infty}\int_{-\infty}^{\infty} r_1\left(\omega_1\omega_2,\Omega_1\Omega_2\right) \\
&\cdot\left[\left\{2\hat{V}_0\left(\omega+\omega_1\right)\hat{V}_0^*\left(\omega+\omega_1+\omega_2\right)\hat{V}_0\left(\omega+\omega_2+\Omega_1\right)\right.\right. \\
&\cdot \hat{V}_0\left(\omega+\omega_2+\Omega_2\right)\hat{V}_0^*\left(\omega+\omega_2+\Omega_1+\Omega_2\right) \\
&-\hat{V}_0\left(\omega+\omega_1\right)\hat{V}_0\left(\omega+\omega_2\right)\hat{V}_0^*\left(\omega+\omega_1+\omega_2+\Omega_1\right) \\
&\left.\cdot\hat{V}_0^*\left(\omega+\omega_1+\omega_2-\Omega_2\right)\hat{V}_0^*\left(\omega+\omega_1+\omega_2+\Omega_1-\Omega_2\right)\right\} \\
&+\alpha\left\{2\hat{U}_0\left(\omega+\omega_1\right)\hat{U}_0^*\left(\omega+\omega_1+\omega_2\right)\hat{U}_0\left(\omega+\omega_2+\Omega_1\right)\right. \\
&\cdot \hat{U}_0\left(\omega+\omega_2+\Omega_2\right)\hat{U}_0^*\left(\omega+\omega_2+\Omega_1+\Omega_2\right) \\
&-\hat{U}_0\left(\omega+\omega_1\right)\hat{U}_0\left(\omega+\omega_2\right)\hat{U}_0^*\left(\omega+\omega_1+\omega_2+\Omega_1\right) \\
&\cdot\hat{U}_0^*\left(\omega+\omega_1+\omega_2-\Omega_2\right) \\
&\left.\left.\cdot\hat{U}_0^*\left(\omega+\omega_1+\omega_2+\Omega_1-\Omega_2\right)\right\}\right] d\omega_1 d\omega_2 d\Omega_1 d\Omega_2
\end{aligned}
\tag{9.77}
$$

Equations (9.68) and (9.69) represent the HO-GTE for the propagation of solitons through birefringent optical fibers.

9.4 DWDM systems

The solitons propagating through a DWDM system can be modeled by the following N-coupled NLSE in the dimensionless form:

$$
iq_z^{(l)} + \frac{D(z)}{2}q_{tt}^{(l)} + g(z)\left\{|q^{(l)}|^2 + \sum_{m\neq l}^{N}\alpha_{lm}|q^{(m)}|^2\right\}q^{(l)} = 0 \tag{9.78}
$$

where $1 \leq l \leq N$. Equation (9.78), once again, models bit-parallel WDM soliton transmission. Here, α_{lm} are known as the XPM coefficients. The fields q_l are expanded in powers of z_a as

$$q_l(\zeta, Z, t) = q_l^{(0)}(\zeta, Z, t) + z_a q_l^{(1)}(\zeta, Z, t) + z_a^2 q_l^{(2)}(\zeta, Z, t) + \cdots \tag{9.79}$$

Equating coefficients of like powers of z_a gives

$$O\left(\frac{1}{z_a}\right): \quad i\frac{\partial q_l^{(0)}}{\partial \zeta} + \frac{\Delta(\zeta)}{2}\frac{\partial^2 q_l^{(0)}}{\partial t^2} = 0 \tag{9.80}$$

$$\begin{aligned} O(1): \quad & i\frac{\partial q_l^{(1)}}{\partial \zeta} + \frac{\Delta(\zeta)}{2}\frac{\partial^2 q_l^{(1)}}{\partial t^2} + \left\{ i\frac{\partial q_l^{(0)}}{\partial Z} + \frac{\delta_a}{2}\frac{\partial^2 q_l^{(0)}}{\partial t^2} \right. \\ & \left. + g(z)\left(|q_l^{(0)}|^2 + \sum_{m\neq l}^{N} \alpha_{lm} |q_m^{(0)}|^2 \right) q_l^{(0)} \right\} = 0 \end{aligned} \tag{9.81}$$

$$\begin{aligned} O(z_a): \quad & i\frac{\partial q_l^{(2)}}{\partial \zeta} + \frac{\Delta(\zeta)}{2}\frac{\partial^2 q_l^{(2)}}{\partial t^2} + \left\{ i\frac{\partial q_l^{(1)}}{\partial Z} + \frac{\delta_a}{2}\frac{\partial^2 q_l^{(1)}}{\partial t^2} \right. \\ & + g(z)\left[2|q_l^{(1)}|^2 q_l^{(1)} + (q_l^{(1)})^2 q_l^{(1)*} \right. \\ & \left.\left. + \sum_{m\neq l}^{N} \alpha_{lm} \{ 2|q_m^{(1)}|^2 q_m^{(1)} + (q_m^{(1)})^2 q_m^{(1)*} \} \right] \right\} = 0 \end{aligned} \tag{9.82}$$

At $O(1/z_a)$, Eq. (9.80), in the Fourier domain, is given by

$$i\frac{\partial \hat{q}_l^{(0)}}{\partial \zeta} - \frac{\omega^2}{2}\Delta(\zeta)\hat{q}_l^{(0)} = 0 \tag{9.83}$$

whose solution is

$$\hat{q}_l^{(0)}(\zeta, Z, \omega) = \hat{Q}_l^{(0)}(Z, \omega)\, e^{-\frac{i\omega^2}{2}C(\zeta)} \tag{9.84}$$

where

$$\hat{Q}_l^{(0)}(Z, \omega) = \hat{q}_l^{(0)}(0, Z, \omega) \tag{9.85}$$

At $O(1)$,Eq.(9.81) is solved in the Fourier domain by substituting the solution given by Eq.(9.84) into Eq.(9.81). This gives

$$\begin{aligned} & i\frac{\partial \hat{q}_l^{(1)}}{\partial \zeta} - \frac{\omega^2}{2}\Delta(\zeta)\hat{q}_l^{(1)} \\ & = -e^{-\frac{i\omega^2}{2}C(\zeta)}\left(\frac{\partial \hat{Q}_l^{(0)}}{\partial Z} - \frac{\omega^2}{2}\delta_a \hat{Q}_l^{(0)} \right) \\ & -g(\zeta)\int_{-\infty}^{\infty} \left(|q_l^{(0)}|^2 + \sum_{m\neq l}^{N} \alpha_{lm} |q_m^{(0)}|^2 \right) q_l^{(0)} e^{i\omega t}\, dt \end{aligned} \tag{9.86}$$

Equation (9.86) is an inhomogeneous equations for $\hat{q}_l^{(1)}$ and with the homogeneous parts having the same structure as in Eq.(9.80). For the non-secularity condition of $\hat{q}_l^{(1)}$, FA gives the condition on $\hat{Q}_l^{(0)}(Z,\omega)$ as

$$\frac{\partial \hat{Q}_l^{(0)}}{\partial Z} - \frac{\omega^2}{2}\delta_a \hat{Q}_l^{(0)} + \int_0^1 \int_{-\infty}^{\infty} e^{\frac{i\omega^2}{2}C(\zeta)} g(\zeta) \cdot \left(|q_l^{(0)}|^2 + \sum_{m\neq l}^{N} \alpha_{lm} |q_m^{(0)}|^2 \right) q_m^{(0)} e^{i\omega t} dt d\zeta = 0 \tag{9.87}$$

Equation (9.87) can be simplified to

$$\frac{\partial \hat{Q}_l^{(0)}}{\partial Z} - \frac{\omega^2}{2}\delta_a \hat{Q}_l^{(0)} + \int_{-\infty}^{\infty} \int_{-\infty}^{\infty} r_0(\omega_1\omega_2) \hat{Q}_l^{(0)}(Z, \omega_1+\omega_2) \cdot \left[\hat{Q}_l^{(0)}(Z,\omega+\omega_1) \hat{Q}_l^{(0)}(Z,\omega+\omega_1+\omega_2) + \sum_{m\neq l}^{N} \alpha_{lm} \hat{Q}_m^{(0)}(Z,\omega+\omega_1) \cdot \hat{Q}_m^{(0)}(Z,\omega+\omega_1+\omega_2) \right] d\omega_1 d\omega_2 = 0 \tag{9.88}$$

Equation (9.88) is the GTE for the propagation of solitons through multiple channels or DWDM systems as seen in Chapter 7. Equation (9.81) will now be solved to obtain $q_l^{(1)}(\zeta, Z, t)$. Substituting $\hat{Q}_l^{(0)}$ into the right-hand side of Eq. (9.86) and using Eqs.(9.84) and (9.87) give

$$\frac{\partial}{\partial \zeta}\left[i\hat{q}_l^{(1)} e^{\frac{i\omega^2}{2}C(\zeta)} \right] = \int_0^1 \int_{-\infty}^{\infty} g(\zeta) e^{\frac{i\omega^2}{2}C(\zeta)} \left(|q_l^{(0)}|^2 + \sum_{m\neq l}^{N} \alpha_{lm}|q_m^{(0)}|^2 \right) q_l^{(0)} e^{i\omega t} dt d\zeta - g(\zeta) e^{\frac{i\omega^2}{2}C(\zeta)} \int_{-\infty}^{\infty} \left(|q_l^{(0)}|^2 + \sum_{m\neq l}^{N} \alpha_{lm}|q_m^{(0)}|^2 \right) q_l^{(0)} e^{i\omega t} dt \tag{9.89}$$

which integrates to

$$i\hat{q}_l^{(1)} e^{\frac{i\omega^2}{2}C(\zeta)} = \hat{Q}_l^{(1)}(Z,\omega) + \zeta \int_0^1 \int_{-\infty}^{\infty} g(\zeta) e^{\frac{i\omega^2}{2}C(\zeta)} \left(|q_l^{(0)}|^2 + \sum_{m\neq l}^{N} \alpha_{lm}|q_m^{(0)}|^2 \right) q_l^{(0)} e^{i\omega t} dt d\zeta - \int_0^{\zeta} \int_{-\infty}^{\infty} g(\zeta') e^{\frac{i\omega^2}{2}C(\zeta')} \left(|q_l^{(0)}|^2 + \sum_{m\neq l}^{N} \alpha_{lm}|q_m^{(0)}|^2 \right) q_l^{(0)} e^{i\omega t} dt d\zeta' \tag{9.90}$$

where

$$\hat{Q}_l^{(1)}(Z,\omega) = i\hat{q}_l^{(1)}(0,Z,\omega)e^{\frac{i\omega^2}{2}C(0)} \tag{9.91}$$

Also, $\hat{Q}_l^{(1)}(Z,\omega)$ is so chosen that

$$\int_0^1 i\hat{q}_l^{(1)} e^{\frac{i\omega^2}{2}C(\zeta)} d\zeta = 0 \tag{9.92}$$

which is going to be an useful relation for subsequent orders. Applying Eq.(9.92) to Eq.(9.90) gives

$$\begin{aligned} &\hat{Q}_l^{(1)}(Z,\omega) \\ &= \int_0^1 \int_0^\zeta \int_{-\infty}^\infty g(\zeta')e^{\frac{i\omega^2}{2}C(\zeta')} \left(|q_l^{(0)}|^2 + \sum_{m \neq l}^N \alpha_{lm} |q_m^{(0)}|^2 \right) q_l^{(0)} e^{i\omega t} dt d\zeta' d\zeta \\ &\quad - \frac{1}{2} \int_0^1 \int_{-\infty}^\infty g(\zeta)e^{\frac{i\omega^2}{2}C(\zeta)} \left(|q_l^{(0)}|^2 + \sum_{m \neq l}^N \alpha_{lm} |q_m^{(0)}|^2 \right) q_l^{(0)} e^{i\omega t} dt d\zeta \end{aligned} \tag{9.93}$$

Now, Eq.(9.90), by virtue of Eq.(9.93), can be written as

$$\begin{aligned} &\hat{q}_l^{(1)}(\zeta,Z,\omega) \\ &= ie^{\frac{i\omega^2}{2}C(\zeta)} \left[\int_0^\zeta \int_{-\infty}^\infty g(\zeta')e^{\frac{i\omega^2}{2}C(\zeta')} \left(|q_l^{(0)}|^2 + \sum_{m \neq l}^N \alpha_{lm} |q_m^{(0)}|^2 \right) \right. \\ &\quad \cdot q_l^{(0)} e^{i\omega t} dt d\zeta' \\ &\quad - \int_0^1 \int_0^\zeta \int_{-\infty}^\infty g(\zeta')e^{\frac{i\omega^2}{2}C(\zeta')} \left(|q_m^{(0)}|^2 + \sum_{m \neq l}^N \alpha_{lm} |q_m^{(0)}|^2 \right) \\ &\quad \cdot q_l^{(0)} e^{i\omega t} dt d\zeta' d\zeta \\ &\quad - \left(\zeta - \frac{1}{2} \right) \int_0^1 \int_{-\infty}^\infty g(\zeta)e^{\frac{i\omega^2}{2}C(\zeta)} \left(|q_l^{(0)}|^2 + \sum_{m \neq l}^N \alpha_{lm} |q_m^{(0)}|^2 \right) \\ &\quad \left. \cdot q_l^{(0)} e^{i\omega t} dt d\zeta \right] \end{aligned} \tag{9.94}$$

which can be further rewritten as

$$\hat{q}_l^{(1)}(\zeta,Z,\omega) = ie^{-\frac{i\omega^2}{2}C(\zeta)} \left[\int_{-\infty}^\infty \int_{-\infty}^\infty \hat{Q}_l^{(0)*}(\omega + \Omega_1 + \Omega_2) \right.$$

$$\cdot \left\{ \hat{Q}_l^{(0)}(\omega + \Omega_1)\, \hat{Q}_l^{(0)}(\omega + \Omega_2) + \sum_{m \neq l}^{N} \alpha_{lm} \hat{Q}_l^{0)}(\omega + \Omega_1)\, \hat{Q}_l^{(0)}(\omega + \Omega_2) \right\}$$

$$\cdot \left\{ \int_0^{\zeta} g(\zeta') e^{i\Omega_1 \Omega_2 C(\zeta')} d\zeta' - \int_0^1 \int_0^{\zeta} g(\zeta') e^{i\Omega_1 \Omega_2 C(\zeta')} d\zeta' d\zeta - \left(\zeta - \frac{1}{2}\right) \int_0^1 g(\zeta) e^{i\Omega_1 \Omega_2 C(\zeta)} d\zeta \right\} d\Omega_1 \Omega_2 \Bigg] \tag{9.95}$$

Thus, at $O(z_a)$,

$$\hat{q}_l(\zeta, Z, \omega) = \hat{q}_l^{(0)}(\zeta, Z, \omega) + z_a \hat{q}_l^{(1)}(\zeta, Z, \omega) \tag{9.96}$$

Moving on to the next order at $O(z_a^2)$, one can note that the GTE given by Eq.(9.88) is allowed to have an additional term of $O(z_a)$ as

$$\frac{\partial \hat{Q}_l^{(0)}}{\partial Z} - \frac{\omega^2}{2} \delta_a \hat{Q}_l^{(0)} + \int_{-\infty}^{\infty} \int_{-\infty}^{\infty} r_0(\omega_1 \omega_2)\, \hat{Q}_l^{(0)}(Z, \omega_1 + \omega_2)$$
$$\cdot \Bigg[\hat{Q}_l^{(0)}(Z, \omega + \omega_1)\, \hat{Q}_l^{(0)}(Z, \omega + \omega_1 + \omega_2) + \sum_{m \neq l}^{N} \alpha_{lm} \hat{Q}_m^{(0)}(Z, \omega + \omega_1)\, \hat{Q}_m^{(0)}(Z, \omega + \omega_1 + \omega_2) \Bigg] d\omega_1 d\omega_2$$
$$= z_a \hat{n}_l(Z, \omega) + O(z_a^2) \tag{9.97}$$

The higher order correction $\hat{n}_l$ can be obtained from the suitable non-secular conditions at $O(z_a^2)$ in Eq.(9.82). Now, Eq. (9.82), in the Fourier domain, is

$$\frac{\partial}{\partial \zeta} \left[i \hat{q}_l^{(2)} e^{\frac{i\omega^2}{2} C(\zeta)} \right] + \hat{n}_l + e^{\frac{i\omega^2}{2} C(\zeta)} \left(i \frac{\partial \hat{q}_l^{(1)}}{\partial Z} - \frac{\omega^2}{2} \delta_a \hat{q}_l^{(1)} \right)$$
$$+ e^{\frac{i\omega^2}{2} C(\zeta)} g(\zeta) \int_{-\infty}^{\infty} \Bigg[2|q_l^{(0)}|^2 q_l^{(1)} + (q_l^{(0)})^2 q_l^{(1)*} + \sum_{m \neq l}^{N} \alpha_{lm} \left\{ 2|q_m^{(0)}|^2 q_m^{(1)} + (q_m^{(0)})^2 q_m^{(1)*} \right\} \Bigg] e^{i\omega t} dt = 0 \tag{9.98}$$

But, again, Eq.(9.92) gives

$$\int_0^1 \hat{q}_l^{(1)} e^{\frac{i\omega^2}{2} C(\zeta)} d\zeta = 0 \tag{9.99}$$

Applying the non-secularity condition (9.99) to Eq.(9.97) gives

$$\hat{n}_l = -\int_0^1 \int_{-\infty}^{\infty} e^{\frac{i\omega^2}{2}C(\zeta)} g(\zeta) \left[2|q_l^{(0)}|^2 q_l^{(1)} + (q_l^{(0)})^2 q_l^{(1)*} + \sum_{m \neq l}^{N} \alpha_{lm} \left\{ 2|q_m^{(0)}|^2 q_m^{(1)} + (q_m^{(0)})^2 q_m^{(1)*} \right\} \right] e^{i\omega t} dt d\zeta \tag{9.100}$$

Using Eqs.(9.84) and (9.94), Eq.(9.100) can be written as

$$\begin{aligned}
\hat{n}_l = &\int_{-\infty}^{\infty}\int_{-\infty}^{\infty}\int_{-\infty}^{\infty}\int_{-\infty}^{\infty} r_1(\omega_1\omega_2, \Omega_1\Omega_2) \\
&\cdot \left[\left\{ 2\hat{Q}_l^{(0)}(\omega+\omega_1)\hat{Q}_l^{(0)*}(\omega+\omega_1+\omega_2)\hat{Q}_l^{(0)}(\omega+\omega_2+\Omega_1) \right.\right. \\
&\cdot \hat{Q}_l^{0)}(\omega+\omega_2+\Omega_2)\hat{Q}_l^{(0)*}(\omega+\omega_2+\Omega_1+\Omega_2) \\
&- \hat{Q}_l^{(0)}(\omega+\omega_1)\hat{Q}_l^{(0)}(\omega+\omega_2)\hat{Q}_l^{(0)*}(\omega+\omega_1+\omega_2+\Omega_1) \\
&\left. \cdot \hat{Q}_l^{(0)*}(\omega+\omega_1+\omega_2-\Omega_2)\hat{Q}_l^{(0)*}(\omega+\omega_1+\omega_2+\Omega_1-\Omega_2) \right\} \\
&+ \sum_{m\neq l}^{N} \alpha_{lm} \left\{ 2\hat{Q}_m^{(0)}(\omega+\omega_1)\hat{Q}_m^{(0)*}(\omega+\omega_1+\omega_2) \right. \\
&\cdot \hat{Q}_m^{(0)}(\omega+\omega_2+\Omega_1)\hat{Q}_m^{(0)}(\omega+\omega_2+\Omega_2) \\
&\cdot \hat{Q}_m^{(0)*}(\omega+\omega_2+\Omega_1+\Omega_2) - \hat{Q}_m^{(0)}(\omega+\omega_1)\hat{Q}_m^{(0)}(\omega+\omega_2) \\
&\cdot \hat{Q}_m^{(0)*}(\omega+\omega_1+\omega_2+\Omega_1)\hat{Q}_m^{(0)*}(\omega+\omega_1+\omega_2-\Omega_2) \\
&\left.\left. \cdot \hat{Q}_m^{(0)*}(\omega+\omega_1+\omega_2+\Omega_1-\Omega_2) \right\} \right] d\omega_1 d\omega_2 d\Omega_1 d\Omega_2
\end{aligned} \tag{9.101}$$

Equation (9.97) represent the HO-GTE for the propagation of solitons through multiple channels or DWDM systems.

References

1. M. J. Ablowitz & H. Segur. *Solitons and the Inverse Scattering Transform*. SIAM. Philadelphia, PA. USA. (1981).
2. M. J. Ablowitz, G. Biondini, S. Chakravarti, R. B. Jenkins & J. R. Sauer. "Four-wave mixing in wavelength-division-multiplexed soliton systems: damping and amplification". *Optics Letters*. Vol 21, No 20, 1646-1648. (1996).
3. M. J. Ablowitz & G. Biondini. "Multiscale pulse dynamics in communication systems with strong dispersion management". *Optics Letters*. Vol 23, No 21, 1668-1670. (1998).
4. M. J. Ablowitz & T. Hirooka. "Resonant nonlinear interactions in strongly dispersion-managed transmission systems". *Optics Letters*. Vol 25, No 24, 1750-1752. (2000).
5. M. J. Ablowitz & T. Hirooka. "Nonlinear effects in quasi-linear dispersion-managed systems". *IEEE Photonics Technology Letters*. Vol 13, No 10, 1082-1084. (2001).
6. M. J. Ablowitz, G. Biondini & E. S. Olson. "Incomplete collisions of wavelength-division multiplexed dispersion-managed solitons". *Journal of Optical Society of America B*. Vol 18, No 3, 577-583. (2001).

7. M. J. Ablowitz & T. Hirooka. "Nonlinear effects in quasilinear dispersion-managed pulse transmission". *IEEE Journal of Photonics Technology Letters.* Vol 26, 1846-1848. (2001).
8. M. J. Ablowitz & T. Hirooka. "Managing nonlinearity in strongly dispersion-managed optical pulse transmission". *Journal of Optical Society of America B.* Vol 19, No 3, 425-439. (2002).
9. M. J. Ablowitz, T. Hirooka & T. Inoue. "Higher-order asymptotic analysis of dispersion-managed transmission systems: solutions and their characteristics". *Journal of Optical Society of America B.* Vol 19, No 12, 2876-2885. (2002).
10. C. D. Ahrens, M. J. Ablowitz, A. Docherty, V. Oleg, V. Sinkin, V. Gregorian & C. R. Menyuk. "Asymptotic analysis of collision-induced timing shifts in return-to-zero quasi-linear systems with pre- and post-dispersion compensation". *Optics Letters.* Vol 30, 2056-2058. (2005).
11. A. Biswas. "Gabitov-Turitsyn equation for solitons in multiple channels". *Journal of Electromagnetic Waves and Applications.* Vol 17, No 11, 1539-1560. (2003).
12. A. Biswas. "Gabitov-Turitsyn equation for solitons in optical fibers". *Journal of Nonlinear Optical Physics and Materials.* Vol 12, No 1, 17-37. (2003).
13. A. Biswas. "Dispersion-Managed solitons in multiple channels". *Journal of Nonlinear Optical Physics and Materials.* Vol 13, No 1, 81-102. (2004).
14. A. Biswas. "Higher-order Gabitov-Turitsyn equation for dispersion-managed vector solitons in birefringent fibers". *International Journal of Theoretical Physics.* Vol 46, No 12, 3339-3354. (2007).
15. A. Biswas. "Higher-order Gabitov-Turitsyn equation for dispersion-managed solitons". *Optik.* Vol 118, No 3, 120-133. (2007).
16. A. Biswas & E. Zerrad. "Higher-order Gabitov-Turitsyn equation for dispersion-managed solitons in multiple channels". *International Journal of Mathematical Analysis.* Vol 1, No 12, 565-582. (2007).
17. I. R. Gabitov & S. K. Turitsyn. "Average pulse dynamics in a cascaded transmission system with passive dispersion compensation". *Optics Letters.* Vol 21, No 5, 327-329. (1996).
18. I. R. Gabitov & S. K. Turitsyn. "Breathing solitons in optical fiber links". *JETP Letters.* Vol 63, No 10, 861-866. (1996).
19. T. Hirooka & A. Hasegawa. "Chirped soliton interaction in strongly dispersion-managed wavelength-division-multiplexing systems". *Optics Letters.* Vol 23, No 10, 768-770. (1998).
20. V. E. Zakharov & S. Wabnitz. *Optical Solitons: Theoretical Challenges and Industrial Perspectives.* Springer, Heidel berg. DE. (1999).

Index

Q

R

S

T

V

W

Nonlinear Physical Science

(*Series Editors: Albert C.J. Luo, Nail H. Ibragimov*)

Nail. H. Ibragimov/ Vladimir. F. Kovalev: Approximate and Renormgroup Symmetries

Abdul-Majid Wazwaz: Partial Differential Equations and Solitary Waves Theory

Albert C.J. Luo: Discontinuous Dynamical Systems on Time-varying Domains

Anjan Biswas/ Daniela Milovic/ Matthew Edwards: Mathematical Theory of Dispersion-Managed Optical Solitons

Hanke,Wolfgang/ Kohn, Florian P.M./ Wiedemann, Meike: Self-organization and Pattern-formation in Neuronal Systems under Conditions of Variable Gravity

Vasily E. Tarasov: Fractional Dynamics

Vladimir V. Uchaikin: Fractional Derivatives in Physics

Albert C.J. Luo: Nonlinear Deformable-Body Dynamics and Waves

Ivo Petras: Fractional Order Nonlinear Systems

Albert C.J. Luo /Valentine Afraimovich (Editors): Hamiltonian Chaos beyond the KAM Theory

Albert C.J. Luo/ Valentine Afraimovich (Editors): Long-range Interaction, Stochasticity and Fractional Dynamics